Published in the United States of America
by Michigan Publishing

DOI: http://doi.org/10.3998/gs

ISBN 978-1-60785-825-6 (paper, issue 2.2)

Front cover image by School of Communication, Hong Kong Baptist University

Secretariat
The Academy of Film
Hong Kong Baptist University
CVA937, Lee Shau Kee Communication and Visual Arts Building, 5 Hereford Road, Kowloon Tong, Kowloon, Hong Kong.
Tel: (852) 3411 7493
Email: gstjournal@hkbu.edu.hk
Website: https://research.hkbu.edu.hk/project/global-storytelling-journal-of-digital-and-moving-images

Supported by
The Academy of Film
The School of Creative Arts
Hong Kong Baptist University

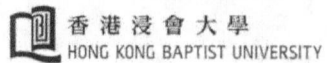

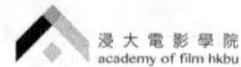

GLOBAL STORYTELLING:
JOURNAL OF DIGITAL AND MOVING IMAGES

Special Issue 2.2 – Narrating Cold Wars
(Winter 2022)

Special Issue Editors: Kenneth Paul Tan & Dorothy Lau

CONTENTS

Letter from the Editor   1
YING ZHU

**Cold War and New Cold War Narratives: Special Issue
Editor's Introduction**   5
KENNETH PAUL TAN

Research Articles

Notes on Cold War Historiography   25
LOUIS MENAND

Tales from the Hot Cold War   37
MARTHA BAYLES

Bomb Archive: The Marshall Islands as Cold War Film Set   57
ILONA JURKONYTĖ

Contents

Das unsichtbare Visier—A 1970s Cold War Intelligence TV
Series as a Fantasy of International and Intranational
Empowerment; or, How East Germany Saved the World
and West Germans Too   87
TARIK CYRIL AMAR

To Whom Have We Been Talking? Naeem Mohaiemen's
Fabulation of a People-to-Come   115
NOIT BANAI

The Man without a Country: British Imperial Nostalgia in
Ferry to Hong Kong (1959)   131
KENNY K. K. NG

Imagining Cooperation: Cold War Aesthetics for a Hot Planet   175
MARINA KANETI

Book Reviews

Through Space and Time
Review of *The Odyssey of Communism: Visual Narratives,
Memory and Culture* edited by Michaela Praisler and
Oana-Celia Gheorghiu, Cambridge Scholars
Publishing, 2021   209
ISABEL GALWEY

Review of *Hollywood in China: Behind the Scenes of the
World's Largest Movie Market* by Ying Zhu, New Press, 2022   217
YONGLI LI

The Cautionary Tale of Painting War Remembrance in
China as a New Nationalism
Review of *China's Good War: How World War II Is Shaping
a New Nationalism* by Rana Mitter, Belknap Press, 2020   225
FUWEI ZUO

Contents

Tracking American Political Currents
Review of *White Identity Politics* by Ashley Jardina,
  Cambridge University Press, 2019, and *Fox Populism:
  Branding Conservatism as Working Class* by Reece Peck,
  Cambridge University Press, 2019   233
  DAVID GURNEY

Contributors   241

Letter from the Editor

YING ZHU

At the end of his introduction that threads together articles in this special collection on the history and histography of the Cold War, our special issue editor Kenneth Paul Tan writes, somewhat wistfully, "It is my hope that the articles in this special issue will contribute valuably to a long but necessary ritual to rid us of that debilitating Cold War specter." In a world where division persists and where audiovisual storytelling thrives in a cold war simulacrum that assigns perpetrators and depicts winners and losers, this might prove to be wishful thinking. We nonetheless hope to move the needle, ever so slightly, at least for the scholarly community, to examine our own precepts and assumptions. Louis Menand, one of our contributors to this special issue and a keynote speaker at the Narrating Cold Wars: An International Interdisciplinary Conference in November 2021, noted in his talk how "the Cold War is given too much power as an explanatory variable," which can be a liability that restricts our interpretive horizons and analytical possibilities.

In my remarks as chair of Menand's keynote speech, I pointed out how reconstructing history is a tricky business, given the elusive nature of the past. How do we "bring a vanished world to life on the page?" to quote Menand's words, given that "the more material you dredge up, the more elusive your subject becomes." From time to time, we are confronted with not just the incompleteness but also the competing versions of the past. History is comprised of not merely facts but also a narrative that assembles and frames facts, a frequently fraught process of omission, elimination, embellishment and at times pure fabrication.

Menand puts it further that "history writing is an imaginative one"—or what our journal calls "storytelling." When it comes to storytelling, historians dredge up the past to make sense of how we get to where we are. The process of digging up the past is, more often than not, purposeful, being guided by our own moral conviction and ideological position as we apply rhetorical persuasion to drive toward closure of some sort, if not a final verdict. The exercise is akin to appraisal or evaluation. To use Stanley Fish's terms, "Evaluation is an act of persuasion rather than demonstration."[1] While cultural critics do their best to persuade, scholars and cultural historians demonstrate meticulously to make a point. To this end, our special issue aims to call attention to our own struggle with the notion of cold war and related scholarship. In the end, the real problem lies perhaps not with the Cold War as too broad a periodization tool in historical summation but "cold war" as a mentality that shapes the way we perceive the world and what divides us and our shared humanities.

The publication of this special issue marks the second anniversary of our journal.[2] As we toddle our way toward year three, we have two more special issues on the horizon. Issue 3.1 will focus on East Asian serial dramas in the era of global streaming services to address aspects of the interplay between local productions and global platforms, including Netflix and Disney+ in the United States and iQiyi in China. Issue 3.2 will zoom in on digital interventions in the public sphere in the broad Asia region, enacted through lighthearted, creative, and resistant communications in multimedia forms (from text and meme to short-video and livestreaming) across platforms and networks. After two years of global trotting, we choose to settle on a

---

1. Stanley Fish, *Is There a Text in This Class? The Authority of Interpretive Communities* (Cambridge, MA: Harvard University Press, 1980), 365–68.
2. Our issue 2.2 also includes a theme-based book review, "Tracking American Political Currents in White Identity Politics and Fox Populism: Branding Conservatism as Working Class" by David Gurney, which covers two books published in 2019 by Cambridge University Press, *White Identity Politics* by Ashley Jardina and *Fox Populism: Branding Conservatism as Working Class* by Reece Peck. The review was supposed to be included in our issue 2.1 but the timing did not work out. We are pleased to include the review in this issue.

particular region for a narrower yet more focused scholarly approach. Asia is naturally our first stop, given the physical base of this journal. It also allows us to leverage local and regional scholarly resources. The focus on digital media in our next two issues meets the demand for attention to storytelling and its ramifications in a fast-evolving digital arena.

Cold War and New Cold War Narratives

Special Issue Editor's Introduction

KENNETH PAUL TAN

Abstract

The historic Cold War, although formally concluded by 1991, continues to widely and to deeply influence, even shape the contours of, the way we think and talk about geopolitics and geoeconomics in the present time. Foreign policy professionals, journalists, scholars, and producers and consumers of popular culture readily turn to tropes, frames, and mental models derived sometimes very literally from this grand-historic episode. Thus, we tend to understand developments in Sino-US relations today, in the first instance at least, through comparisons with the intense superpower rivalry between the United States and the Soviet Union in the bipolar world of the Cold War. By referring to the articles in this special issue on narrating cold wars, its guest editor describes how such frames, models, and mentalities, as they are realized in and conveyed through narratives, can be challenged in a variety of ways.

Keywords: The Cold War and New Cold War, Sino-US relations, narratives, narration, historiography, Cold War themed films, television series

It is 2022. The historic Cold War ended just over three decades ago. But its fascination for people all over the world continues, not only among foreign policy professionals, whose grand-strategic perspectives on geopolitics and geoeconomics are still often anchored to Cold War mentalities and the narratives through which they are vividly expressed but also in journalism, academia, as well as popular culture and the arts. This special issue of *Global Storytelling* centers on the narratives and narration that have featured in the ways that people during the Cold War made sense of their world for themselves and for others, in the ways that people today try to understand the Cold War of the past, and in the ways that people today think about their contemporary world and where it might be headed.

When Mikhail Gorbachev, the Soviet Union's last president before it collapsed in 1991, died on August 30, 2022, news media from all over the world put out extensive obituaries of the reformist Soviet leader under whose colorful rule the USSR crumbled and the Cold War ended bloodlessly. There was no shortage in the news media of reports carrying soaring tributes from the global elite; some reports were more nuanced, but all recognized his world-historic role at the "end of History."[1] Russian president Vladimir Putin, however, blamed Gorbachev for the failure to prevent the fall of the Soviet empire, pointing out that most of this high praise had come through the Western media.[2] China's coverage of Gorbachev's death was noticeably "low-key," indicating the Chinese leadership's aversion to political reform and cautious attitude toward economic liberalization. The lesson learned from the historic role Gorbachev's *perestroika* and *glasnost* had played in the USSR's fall affirms a fast-rising China's commitment today, enforced by the strong hand of its president Xi Jinping, to asserting, even tightening, political and economic control.[3]

1. Just within a couple of days of Gorbachev's death, for instance, Archie Bland, "The death of Mikhail Gorbachev"; and Martin Farrer, "Mikhail Gorbachev: Tributes Pour in for 'One of a Kind' Soviet Leader," *Guardian*, August 31, 2022; Marilyn Berger, "Mikhail S. Gorbachev, Reformist Soviet Leader, Is Dead at 91," *New York Times*, August 30, 2022; " 'One of a Kind': World Reactions to Death of Mikhail Gorbachev," *Aljazeera*, August 31, 2022.
2. Kevin Liffey, "West Mourns Gorbachev the Peacemaker, Russia Recalls His Failures," Reuters, September 1, 2022.
3. Mimi Lau and Guo Rui, "China Offers Tributes to late Soviet Leader Mikhail Gorbachev for Fostering Beijing-Moscow Ties," *South China Morning Post*, August 31, 2021.

The Western and Eastern Blocs, and the hard political and ideological line dividing them, no longer exist formally thirty years after the end of the Cold War, but the global elite's attitudes and behaviors reveal that their ghostly presence haunts us today. Current events, it would seem, continue in large part to lie in the shadow of Cold War geopolitical division, binary and polarizing logic, and rhetorical styles. In this last decade especially, news reports and commentaries often reference the Cold War in analysis of current geopolitics and geoeconomics, particularly in trying to make sense of Sino-US rivalry. We are—as many would point out, and some without reservation—at the start of a New Cold War, essentially the same script and same narrative, but some of the roles are played by different actors, wearing different costumes. British historian Niall Ferguson, for instance, describes a "Cold War II," where "the roles [between China and the original Soviet Union] have been reversed. China is now the giant, Russia the mean little sidekick. China under Xi remains strikingly faithful to the doctrine of Marx and Lenin. Russia under Putin has reverted to Tsarism."[4]

Historiography

Today, the historic Cold War features prominently in undergraduate syllabuses and graduate dissertations in the fields of history, international relations, political science, area studies, as well as cultural, media, and film studies. Scholars continue to engage in lively debate over the details, evidence, concepts, methods, assumptions, theories, literatures, predictions, policy recommendations, and academic gatekeeping that shape and are in turn shaped by academic studies of the Cold War. Almost twenty years ago, American sociologist Craig Calhoun described how the field of Cold War studies was dominated by two approaches. A "traditional" approach, favored

---

4. Niall Ferguson, "Cold War II Has America at a Disadvantage as China Courts Russia," *Boston Globe*, January 20, 2020.

by the US establishment, placed the responsibility for the Cold War on an aggressively expansionist Soviet Union eager to advance worldwide Communist revolution. According to this orthodox approach, it was the Soviet Union's provocative actions that forced the United States out of isolationism to contain but also to counter Communist insurgencies around the world, including through the reconstruction of its allies' postwar economies via a massive foreign aid plan. Challenging this official US version was a "revisionist" approach that shifted responsibility for the Cold War onto the United States. This more critical view emerged during the Vietnam War at a time of rising public skepticism—at home and abroad—of America's role in world affairs.[5]

A third approach, which Calhoun briefly described then as an emerging "post-revisionist" approach, has today become more widespread. Much less interested in assigning blame, post-revisionist scholarship appears more balanced, nuanced, and complex. It is more concerned with rigorous contextualization and rethinking of all relevant parties' motivations, perspectives, actions, and accomplishments during the period. *The Free World: Art and Thought in the Cold War*, authored by Harvard English professor Louis Menand, is exemplary in this respect.[6] In his review of Menand's book about American artistic and intellectual achievements during the Cold War, Stanford professor Mark Greif identified its central argument in this way: "Artistic success owes little to vision and purpose, more to self-promotion, but most to unanticipated adoption by bigger systems with other aims, principally oriented toward money, political advantage, or commercial churn. For the greatness and inevitability of artistic consecration, Menand substitutes the arbitrary confluences of forces at any given moment."[7]

---

5. Craig Calhoun, "Cold War," in *Dictionary of the Social Sciences*, ed. Calhoun (Oxford University Press, 2002).
6. Louis Menand, *The Free World: Art and Thought in the Cold War* (Farrar, Straus and Giroux, 2021).
7. Mark Greif, "The Opportunists," *Atlantic*, June 2021.

This special issue of *Global Storytelling*, assembled around the theme of narrating cold wars, begins with an article by Menand. In it, he reflects on the historiographical considerations involved in the decade-long research and writing process leading to the publication of his critically acclaimed *The Free World*. These include, among others, establishing the appropriate time frames for analysis; determining the meanings and implications of big ideas like "freedom" for different parties in different contexts; uncovering the consequences of faulty evidence, misinterpretation, and misattribution in well-established scholarship; and decentering conventional reasons offered to explain artistic achievement. All the while, Menand recognizes the challenge of self-consciously maintaining an awareness of his own subjectivity as a product of the history he was trying to write.

An article on historiography of this kind clearly belongs to this special issue on narrating cold wars because history writing most often requires establishing narrative linkages so that available facts of the past, once verified, can be configured to tell a story that is continuous, logical, coherent, plausible, comprehensible, transmissible, and significant. Since ancient times, history (and some philosophy) has been a form of storytelling. Classic history tends to follow typical plots found in literary genres.[8] Traditional and revisionist accounts of the Cold War often narrate the past by organizing materials into adversarial relationships between protagonists and antagonists, each with destinies to fulfill and problems to overcome, and the motivation, potential, and means to do it. They have become the grand narratives of the Cold War, through which historical explanations, interpretations, and criticisms of not only the past but also the present and even the future might be enabled but also, at the same time, constrained.

Metanarratives of history attempt to tell us something even more abstract than historical grand narratives. They try to reveal the logic, shape, direction, and even purpose of History (with a capital "H"), connecting

---

8. Hayden White, *Metahistory: The Historical Imagination in Nineteenth-Century Europe* (Baltimore: Johns Hopkins University Press, 1973).

past, present, and future in one dynamic and meaningful trajectory. For instance, American political scientist Francis Fukuyama famously wrote *The End of History and the Last Man* at the momentous end of the Cold War, arguing triumphantly that History was coming to an end since capitalist liberal democracy seemed to have triumphed over Fascism and Communism.[9] In the historical battle of ideologies, all political communities in this diverse world would transition into capitalist liberal democracies according to their own path and pace, at which stage all human and social needs will be met through the practical and collective application of reason and there will be equal and reciprocal recognition of everyone's right to freedom. Fukuyama's metanarrative of history, built upon Platonic and Hegelian philosophical foundations to arrive at what seem obviously to be liberal conclusions, was in many respects a renewed version of modernization theory, dominant though already controversial in the 1950s and 1960s. It was a theory—made intelligible, persuasive, and exciting through a metanarrative of human enlightenment and progress—that linked economic growth and development, urbanization, industrialization, and mass education with the rise of a middle class and a transition from "traditional" authoritarian to "modern" democratic systems.[10]

Just as histories often find their most meaningful and emotionally satisfying expression through narrative and narration, so too do philosophies of history through metanarratives and metanarration. Fukuyama's philosophy of history in 1992 intersected with momentous world events in the years leading up to it, to create a declaratory moment in which celebration and hyperbole seemed appropriate. In the decades that followed this declaration, it has become all too clear that History is nowhere close to ending. In established liberal democracies, poverty remains persistent, inequality has

---

9. Francis Fukuyama, *The End of History and the Last Man* (New York: Free Press, 1992).

10. Seymour Martin Lipset, *Political Man: The Social Basis of Politics* (New York: Doubleday, 1960), contested most recently by Daron Acemoglu and James A. Robinson, "Non-modernization: Power–Culture Trajectories and the Dynamics of Political Institutions," *Annual Review of Political Science* 25, no. 1 (2022): 323–39.

deepened along many different axes, and corporate capital increasingly controls the levers of power in ways that reduce democracy to a mere sham, as capitalism grows mostly unfettered. Right-wing populism has risen in both authoritarian and democratic countries. Autocratization is now being reported as a rising trend around the world.[11] And the rise of China has buoyed talk about an "Asian century"—even a "Chinese century"—poised to replace Pax Americana with Pax Sinica.[12] Without the emergence of more satisfactory narratives and metanarratives to explain our still-tumultuous history, the Cold War narratives that brought so much meaning (as well as violence and turmoil) to ordinary people's lives and livelihoods from 1945 to 1991 will persist and continue to prevail, endangering all corners of the world with a constant threat of devastating war and the inability to cooperate internationally in the urgent pursuit of implementable solutions for global problems of the most existentially critical kind.

Popular Culture and the Arts

One powerful way in which Cold War narratives persist and prevail is through their production, circulation, consumption, and reappropriation in popular culture and the arts. Cinema and television continue to feature Cold War themes, character types, settings, and styles for entertaining today's audiences. For example, in *The Odyssey of Communism: Visual Narratives, Memory and Culture*,[13] a collection of essays edited by Michaela Praisler and Oana-Celia Gheorghiu, both academics at the University of Galați, one finds interdisciplinary analysis of films that present Communism in terms

---

11. V-Dem Institute, *Autocratization Turns Viral: Democracy Report 2021* (Gothenburg: University of Gothenburg, 2021).

12. Hal Brands, "The Chinese Century?" *National Interest*, February 19, 2018; William Rees-Mogg, "This Is the Chinese Century," *(London) Times*, January 3, 2005.

13. Michaela Praisler and Oana-Celia Gheorghiu, eds., *The Odyssey of Communism: Visual Narratives, Memory and Culture* (Newcastle upon Tyne: Cambridge Scholars Publishing, 2021).

that range from the most hellish to the rosiest. The book, which suggests that an insufficiently interrogated Communist past demands revisiting, is reviewed by Isabel Galwey in this issue.

Over the last five years alone, there have been numerous critically or commercially successful films as well as television drama and comedy series that are variations on Cold War themes and narratives. Many of them are popularly available on streaming platforms like Netflix and Apple TV+. A broad-ranging sample of such shows might include:

- *Milada* (2017, directed by David Mrnka, Czech Republic/United States): A biographical film about Milada Horáková, who struggled as a political activist in Czechoslovakia to advance freedom, democracy, and human rights, opposing first the Nazis and then the Communist Party. She was executed under fabricated charges. The film ends with the caption: "Over 2 billion people live under dictatorships around the world today. We dedicate this movie to their fight for freedom."
- *A Very Secret Service* (season 1 in 2015, season 2 in 2018, directed by Alexandre Courtès, France, Netflix): A comedy series, full of parody and wit, about André Merlaux, who is recruited into the French Secret Service during the height of the Cold War. He undergoes training and undertakes outlandish missions while, in the background, France is confronted abroad by decolonization and by countercultural developments at home.
- *Cold War* (2018, directed by Paweł Pawlikowski, Poland/United Kingdom/France/Belgium): A bleak arthouse film about a doomed pair of lovers, Wiktor and Zula, who are a poor match but unavoidably attracted to each other. Over the 1940s to the 1960s, their story moves from Poland, to Germany, to Yugoslavia, to France, and then back again to Poland, where—at the end of the film—they attempt suicide. The background is saturated with Cold War politics, in many ways responsible for making their already troubled union impossible and their happiness unattainable.

- *Traitors* (2019, created by Bathsheba Doran, United Kingdom, Netflix): A television drama miniseries set in postwar London, where a brilliant young woman, Feef Symonds, agrees to spy on her own government for the Americans, who convince her that there is a Soviet spy in the UK Cabinet Office.
- *For All Mankind* (three seasons since 2019, created by Ronald D. Moore, Matt Wolpert, and Ben Nedivi, United States, Apple TV+): An "alternate history" television drama series exploring what could have happened—from the 1960s to the 1990s—if the Soviet Union were, counterfactually, to have put a man on the moon ahead of the United States and if the "space race" never ended.
- *Operation Buffalo* (2020, directed by Peter Duncan, Australia, Netflix): A satirical comedy-drama television series based on British nuclear bomb tests (code named Operation Buffalo) carried out from a secret military base in the remote outback of Australia in the 1950s. A caption at the beginning of each episode reads: "This is a work of historical fiction, but a lot of the really bad history actually happened."
- *Da 5 Bloods* (2020, directed by Spike Lee, United States, Netflix): A film about a group of four African American Vietnam War veterans, who return to Vietnam in their old age. Their objective is to locate and bring home to the United States the remains of their squad leader, who had fallen in battle. They also aim to retrieve a chest of gold bars that they had found and buried while on duty there.
- *Gloria* (2021, directed by Tiago Guedes, Portugal, Netflix): A television thriller series about a young man in the Portuguese village of Glória do Ribatejo, João Vidal, whom the Soviet Union recruits as a spy to undertake risky missions at the height of the Cold War.
- *Autumn Girl* (2021, directed by Katarzyna Klimkiewicz, Poland, Netflix): A musical film about Kalina Jedrusik, a singer and actress in 1960s Poland, who resolves to maintain her free-spirited lifestyle and sex-symbol image in a socially and culturally conservative Communist

society, where men in power use their access to state levers to advance their sexual desires over attractive women like Jedrusik.

- *18½* (2021, directed by Dan Mirvish, United States): A dark comedy film about a White House transcriber, Connie, who finds the only copy of the missing eighteen and a half minutes of President Richard Nixon's politically scandalous tapes and, together with a journalist Paul, tries in vain to turn over the information.
- *Stasikomödie* (A Stasi comedy) (2022, directed by Leander Haußmann, Germany): A comedy film about famous author Ludger Fuchs who, as a young man in the 1980s, was recruited by the East German Secret Police (the Stasi), and then deployed to spy on subversives living in a bohemian part of Berlin. In their company, he eventually became a talented poet and then the face of resistance to the Communist German Democratic Republic.
- *Kleo* (2022, created by Hanno Hackfort, Bob Konrad, and Richard Kropf, Germany, Netflix): A television drama series, with dark comedy elements, about Kleo Straub, a quirky and highly skilled former agent of the East German Stasi, who—after the fall of the Berlin Wall—is released from prison and takes "Tarantino-style" revenge on those who betrayed her.

Out of this broad-ranging sample, a few of the shows have clearly been constructed literally from old Cold War narratives, reinforcing them in many ways. Others "replay" these narratives with highly self-conscious irony or a parodic sensibility that bestows a pleasurable (and perhaps even eye-opening) strangeness upon the Cold War tropes that have become so normalized in everyday life. Others focus on the tragic elements of Cold War narratives, as a direct and perhaps cathartic mode of critique. With stories set in the United States, the United Kingdom, France, Germany (East and West), Czechoslovakia, Poland, Australia, and Portugal, several of these shows reveal not only the intricately global effects of the historic Cold War, but also how globally entertaining these Cold War narratives have been and can be to audiences everywhere.

Also interesting to note is the presence of strong female protagonists in many of these recent shows: a charismatic activist in *Milada*, who is executed for her commitment to freedom, democracy, and human rights in Communist Czechoslovakia; a skillful and, at first, thoroughly loyal agent of the Stasi in *Kleo*, who takes revenge on the people who betrayed her, including those within the Communist system; a sexually liberated performer in *Autumn Girl*, who asserts her independence, creativity, and free-spiritedness in Communist Poland; a British spy in *Traitors*, working for the United States to identify a Soviet spy in the British political establishment; a White House transcriber in *18½*, who almost manages to expose the missing eighteen and a half minutes of Nixon's infamous tapes; and heroic female US astronauts in *For All Mankind*. While none of the Cold War–era films and television shows discussed in the articles of this special issue feature strong female protagonists, it may be a stretch to suggest that these more recent shows that do indicate an emergence in popular culture of feminist modes of resistance to a Cold War patriarchy—after all, some of them end up reinforcing Cold War masculinities. But they do nevertheless present avenues for reimagining the affective power of narratives to reshape conditions of possibility for a truly post–Cold War world.

The United States and the Communist Bloc

The first three articles in this special issue on narrating cold wars revolve around the United States. As discussed earlier, Louis Menand's article, "Notes on Cold War Historiography," reflects on the historiographical considerations behind the writing of his *The Free World*, a book that challenged academic orthodoxies surrounding American artistic and intellectual achievements in the "cultural Cold War."

In her article, "Tales from the Hot Cold War," Martha Bayles, a humanities professor at Boston College, analyzes three Korean War films and thirteen Vietnam War films—that is to say, films about the proxy wars in Asia—produced from the 1950s to the 1980s by what she calls the "Washington-Hollywood pact." She discusses the implications of this evolving partnership between the US political and foreign policy establishment and the American movie industry, which she says behaves like "an old married couple who quarrel at home but are deeply united in their outward-facing dealings with the world." This pact was responsible for some of the most vivid and even inspiring cinematic storytelling to justify war, at home and abroad. However, Bayles argues that this pact is, today, in dire need of repair, as Hollywood seems more willing to compromise its independence for a bigger—though illusory—share of the global, especially the Chinese, market.

The topic of Hollywood's relationship with China's gigantic movie industry is also the subject of *Hollywood in China: Behind the Scenes of the World's Largest Movie Market*, a book by Ying Zhu, who is a Hong Kong Baptist University film professor and Professor Emeritus at the City University of New York.[14] Zhu's nuanced depiction of the Sino-Hollywood dynamic is a post-revisionist treatment of the Cold War and its cultural legacies. The book is reviewed by Yongli Li in this issue.

In her article, "Bomb Archive: The Marshall Islands as Cold War Film Set," film curator and researcher Ilona Jurkonytė discusses how the logic, modality, and resources of Hollywood were extended to the urgent task of audio-visual documentation of American nuclear bomb testing within the coral reef in the Marshall Islands commonly known as Bikini Atoll (though Jurkonytė uses the Marshallese transliteration "Pikinni"). Jurkonytė explains how a "bomb archive" was thereby produced and then used to construct "scientific" and "technological" narratives to justify what she argues were in fact efforts by the United States to expand its direct influence into oceanic spheres. Thus, she argues, "the production and circulation of the bomb archive is at the core of nuclear colonial injustices," in which Marshall Islanders were subjected to relocation, resettlement, and dangerous levels of

---

14. Ying Zhu, *Hollywood in China: Behind the Scenes of the World's Largest Movie Market* (New York: New Press, 2022).

radioactive exposure. Jurkonytė provocatively urges the reader to view the Marshall Islands not just analogically but literally, as a film set to produce archival materials that have gone into the global weaving of stories that obscure what she argues is, in fact, neocolonial expansion.

If the first three articles question—in their own different ways—the (popular) cultural achievements of Cold War America, the fourth article in this special issue critiques a popular East German television series, *Das unsichtbare Visier*, as propagandistic storytelling. Tarik Cyril Amar's article is cheekily titled "*Das unsichtbare Visier*—A 1970s Cold War Intelligence TV Series as a Fantasy of International and Intranational Empowerment; or, How East Germany Saved the World and West Germans Too." In it, the history professor at Koç University discusses how the Cold War has engendered a type of character in popular culture—the intelligence agent, fictional and thoroughly heroic. Best known among them is the United Kingdom's Agent 007, James Bond, who has not only survived but thrived as a movie franchise well beyond the original Cold War. In East Germany were the Stasi agents featured in *Das unsichtbare Visier*, a series that was televised during the 1970s. These agents were stylized as cosmopolitan sophisticates, moving around the world to save innocent people, in both the East and the West, from the evils of capitalism (but also from neo-Nazism and neo-Fascism) and from its ruthless elite. *Das unsichtbare Visier* is an elaborate fantasy of Communist virtue, sophistication, and heroism, disproportionate to East Germany's stature in the Cold War world. As Amar notes, "Like Britain's James Bond, these were agents of an at best middling power doing major things in the world at large."

Nonalignment

The first four articles in this special issue feature principal participants of the historic Cold War: the US superpower helming the capitalist-liberal-democratic West (the so-called First World) on the one hand and, on the other,

East Germany, a key member of the Soviet-led Communist bloc (the Second World). In between these two blocs were numerous countries—mainly newly independent nations of the so-called Third World—that chose to be politically and ideologically "nonaligned," forming an international movement to express and defend this position.

The Non-Aligned Movement (NAM) is the central subject of the fifth article in this special issue, titled "To Whom Have We Been Talking? Naeem Mohaiemen's Fabulation of a People-to-Come." Its author, Noit Banai, is an art historian and professor at Hong Kong Baptist University. Her article is a critical analysis of Bangladeshi filmmaker and Columbia University professor Naeem Mohaiemen's 2017 film installation *Two Meetings and a Funeral* and his 2019 and 2021 performance-lecture *The Shortest Speech*. Banai's article locates in Mohaiemen's work the difficult question of whether NAM was ever able to surpass or even transcend the colonial Western models of society and state that had been so deeply imprinted in the former colonies and the mindsets of its nationalist elite. Banai notes how Mohaiemen's film uneasily juxtaposes two highly contrasting registers: the visionary speeches given by these towering nationalist leaders at the early NAM conferences and the "fragmented streams of consciousness about a miscarried past" characterizing the voices of present-day commentators whom the film engages to make sense of that strangely heroic past. The speeches at these NAM conferences, captured and disseminated on television, were—Banai explains—a form of "fabulation," as Gilles Deleuze uses the term. They were an act of "legending," a minority's counter-narration to create a people-to-come.

The sixth article in this issue explores a different kind of nonalignment. The British colonial administration in Hong Kong had put in place, for the sake of securing political stability, an active practice of film censorship that would give their colony the appearance of neutrality toward the United States (with whom Britain had a special relationship) and US-backed Taiwan on the one hand and Communist China (whose subversive activities Britain was highly apprehensive about) on the other. In "The Man Without a Country: British Imperial Nostalgia in *Ferry to Hong Kong* (1959)," Kenny K. K. Ng, a film professor at Hong Kong Baptist University, closely analyzes the star-studded but commercially and critically unsuccessful British film *Ferry to Hong Kong*, which was also a sort of British-American "joint venture." As Ng argues, "*Ferry to Hong Kong* is taken as an imaginary battleground in Britain's ideological war against the forewarned Communist intrusions into the Crown colony." Just as it was constrained by the considerations of Cold War politics, the film also reflected a deep sense of imperial nostalgia and of loss for the British empire, confronted at the time by pressures for decolonization and a decline in its international prestige and influence.

The New Cold War

While the first six articles in this special issue focus mainly on events that happened during the Cold War period, its final article considers the legacies of the Cold War for geopolitics today. Marina Kaneti, an international affairs professor at the National University of Singapore, examines the possibility of international cooperation to tackle the climate crisis today, in the context of a "New Cold War" between the United States and China. In her article, "Imagining Cooperation: Cold War Aesthetics for a Hot Planet," Kaneti performs a visual analysis of popular magazine covers in the United States and the Soviet Union, noting how the public today remembers the space race during the Cold War principally through the lens of "competition," even though there was substantial cooperation at the time. Drawing insights from the politics of aesthetics, Kaneti considers the affective power of images to orientate/disorientate the common sense basis of legitimating collaborative action to deal with the climate crisis, even, as she describes, "in times of extreme hostility and ideological opposition."

Indeed, extreme hostility is by no means a hyperbolic description of Sino-US relations in the present time. Harvard political scientist Graham Alison theorized that a rising China that threatens to displace US global hegemony in a unipolar, post–Cold War world will create enough stress in the international system to increase significantly the likelihood of war. This much-cited "Thucydides trap" was based on Alison's study of sixteen historically similar case studies, out of which twelve had led to war.[15] Today, there are a number of provocative issues that could further heighten these chances of war, such as territorial disputes in the South China Sea; human rights issues in Tibet and Xinjiang; tense relations with North Korea, Hong Kong, and Taiwan; cyber espionage; and trade and tech wars.

The narrative of China's rise begins in the late 1970s, when major economic reforms led to continuous economic growth at a consistently impressive level. Even if we recognize the many weaknesses of China's economy today, it is still predicted to overtake that of the United States to become the world's largest this century. Under President Xi, many have observed a more nationalistic China, its "wolf warrior diplomacy" appearing to replace former assurances of a "peaceful rise." The "wolf warrior" term comes from a series of commercially successful Chinese patriotic action films. The tagline of one of these films was, "Even though a thousand miles away, anyone who affronts China will pay."[16] Oxford University history professor Rana Mitter finds a connection between China's "new nationalism" and a shift in the state's direction of war narratives and memories: its war against Japan during World War II, once viewed through a narrative of victimization, has now become a founding myth. His book *China's Good War: How World War II Is Shaping a New Nationalism* is reviewed by Zuo Fuwei in this issue.[17]

Meanwhile, the narrative of American decline and its struggle to retain its preeminent status in the world includes pessimistic accounts of economic stagnation, unaffordable military commitments around the world, the rise of authoritarian populism, cultural-intellectual exhaustion, institutional decay,

---

15. Graham Alison, *Destined for War: Can America and China Escape Thucydides's Trap?* (New York: Houghton Mifflin Harcourt, 2017).

16. Ben Westcott and Steven Jiang, "China Is Embracing a New Brand of Foreign Policy. Here's What Wolf Warrior Diplomacy Means," CNN, May 29, 2020.

17. Rana Mitter, *China's Good War: How World War II Is Shaping a New Nationalism* (Cambridge, MA, and London: Belknap Press, 2020).

and dysfunctional domestic politics. America has become a victim of its own success.[18] Its former president Donald Trump has often named China as the number one threat. Continuing that approach, current president Joe Biden is determined to do everything he can to prevent China from becoming number one.[19] In a public appearance in July 2022, the Federal Bureau of Investigation (FBI) director described China as the "biggest long-term threat to our economic and national security," citing examples that included massive-scale cyber espionage, technology theft, interference in domestic politics and elections, and a possible invasion of Taiwan. China's Foreign Ministry spokesman warned him not to "hype up the China threat theory" with "irresponsible" remarks to "smear and attack China," also advising him to "cast away imagined demons" and his "Cold War mentality."[20]

What we now have is the beginnings of a new and already very tense superpower rivalry between a declining United States and an emerging China, the start of a New Cold War that is likely to continue spawning Cold War–style narratives that will repolarize the world, with either/or ways of thinking to mobilize for "us" and against "them," often through the demonization of "them" into an "enemy-other." In these times of extreme hostility, can new and affectively powerful narratives expressed through popular culture (films, television, visual art, etc.) reorientate the mindsets of peoples of democratic and less democratic countries to demand that their leaders pay more attention to those problems that legitimately and urgently require global cooperation to solve? These problems include environmental degradation, poverty, inequality, social justice, and several other issues identified, for example, in the United Nations' Sustainable Development Goals.[21]

---

18. For example, Ross Douthat, *The Decadent Society: How We Became the Victims of Our Own Success* (New York: Avid Reader, 2020).
19. "Joe Biden Is Determined That China Should Not Displace America," *Economist*, July 17, 2021.
20. Gordon Corera, "China: MI5 and FBI Heads Warn of 'Immense' Threat," BBC News, July 7, 2022.
21. Kenneth Paul Tan, *Movies to Save Our World: Imagining Poverty, Inequality, and Environmental Destruction in the 21st Century* (Singapore: Penguin, 2022).

Narrating a Cold War Conference

The articles in this special issue were selected from more than sixty papers presented at a three-day conference on narrating Cold Wars, which ran from November 11–13, 2021, and was organized by Hong Kong Baptist University's School of Communication (and Film), in collaboration with the Academy of Visual Arts and the Department of Government and International Studies. I had the pleasure of curating the conference and would like to place on record my thanks to members of the organizing committee: Noit Banai, Jean-Pierre Cabestan, Alistair Cole, Cherian George, Mateja Kovacic, Daya Thussu, and Ying Zhu. Videos of all conference sessions can be viewed at https://www.hkbu.online/narratingcoldwars/.

In my opening remarks at the conference, I asked, "Is the Cold War an old specter that tenaciously haunts our world, begging to be exorcized and therefore liberated from our collective lack of imagination? If so, then we might want to think of these three days as an exorcism, a ritual of diverse liturgies to summon the ghosts of the Cold War so that we may set them, and us, free at last." It is my hope that the articles in this special issue will contribute valuably to a long but necessary ritual to rid us of that debilitating Cold War specter.

Research Articles

Notes on Cold War Historiography

LOUIS MENAND

Abstract

In this essay, Menand raises historiographical questions about the Cold War, arising mainly from his experience of authoring *The Free World: Art and Thought in the Cold War* (New York: Farrar, Straus and Giroux, 2021).

Keywords: Cultural Cold War, historiography, American intellectual and artistic achievements, US Cold War politics

I spent ten years writing a book called *The Free World: Art and Thought in the Cold War*, a history of the period from 1945 to 1965. I was not trained as a historian; my field is English literature, and my book is about art and ideas. But writing history teaches you things about writing history: you run into conceptual and practical problems, and you have to come up with solutions. This paper is an informal look at some of the historiographic issues, as I encountered them, in Cold War studies.

It can seem that the first thing a historian needs to do is establish periodization. When we use the term *Cold War*, what stretch of time are we referring to? But periodization always comes second. It depends on a prior act of interpretation. We need to have already decided what the Cold War was, or was about, before we set before-and-after dates to it. And that decision is a function of point of view.

If we interpret the Cold War as the name for US-Soviet relations (which is how George Kennan interpreted it), the dates would be 1917 to 1991.

https://doi.org/10.3998/gs.3427

Relations warmed and cooled in those years, but at no time did either nation not consider the other to be potentially a threat to its own interests. The customary dates of the Cold War, 1947 (the year of the Truman Doctrine) to 1991 (when the Soviet Union voted itself out of existence), pick out the period when the effects of US-Soviet relations became global, and when the struggle boiled down to what was essentially an arms race.

Ideologically, the Cold War names a much shorter period (in fact, the period of my own book, 1945–1965), and that is the most common use of the term. Phrases like "Cold War liberalism" or "the Cold War university" are usually taken to refer to that period. In domestic politics, the Cold War is associated with McCarthyism and anti-Communist crusades, and that period is even shorter, from 1947, when President Truman initiated a loyalty program for federal employees and Congress launched its investigation of Communists in Hollywood, to 1957, when the Supreme Court handed down a series of decisions restricting the power of government to prosecute individuals for their political beliefs.

After 1957, the Cold War receded as a political issue in the United States, although "international Communism" continued to dominate foreign policy, culminating in the American military intervention in Vietnam in 1965. The (short-term) failure of that intervention reconfigured American foreign policy and removed Communism from the front burner in domestic politics.

These are all US-centric perspectives, however. From a European point of view, the Cold War could be thought of as a civil war, capitalism versus socialism, with roots in the nineteenth century. This "civil war" began a distinctive chapter in 1945, or even 1944, when the Soviet Union's designs on Eastern Europe became clear. For Eastern Europeans, this Cold War did not end until 1989, the year of the Velvet Revolution and the toppling of the Berlin Wall. For Western Europeans, the threat of Communism was largely over after 1956, when the Red Army suppressed the Hungarian Revolution and the Soviet Union lost support among most Western activists and intellectuals.

My own conclusion after dealing with periodization problems is that useful historical time frames are actually quite short, from three to five years.

Beyond that, you cannot hold circumstances constant sufficiently to permit generalization. When an individual figure, an artist or political actor, becomes iconic, it is almost always because of work accomplished or activities undertaken in a period lasting a few years. There is an intersection of an individual life history and social forces that make a new kind of thing, a book or an artwork or a political movement, possible. Then, forces shift, creating conditions for the possibility of something else, the iconic figure becoming one of those conditions.

I'm not convinced that we should generalize about historical periods at all. About the only thing I found to be consistent in the years 1945 to 1965 is that everyone used the language of freedom (hence the title of my book). But what people used that language to justify and what they meant by "freedom" were so various that the concept reduces only to something like "anti-totalitarian." "Freedom" meant, relative to authoritarian regimes and command economies, a lack of coercion by collectivities—in particular, by the state. But the concept encompassed free markets as well as freedom of speech, states' rights as well as civil rights, and not every proponent of one was a supporter of the other.

If we start our story in 1945, we want to be careful not to oversimplify the ideological situation. As a practical matter, Truman's policy of committing the United States to intervene whenever a democratic government was endangered anywhere in the world was untenable. The United States did nothing to prevent—and, later on, did nothing to undo—the Soviet colonization of Eastern Europe. Kennan's policy of containment was anti-interventionist. It was to keep the Communists in their box. What went on on the other side of the Iron Curtain stayed on the other side of the Iron Curtain.

But, in principle, simply the existence of totalitarian states is an affront to democratic values. Totalitarian governments throw their political opponents into prison and kill them; they pursue genocidal policies toward their own people; they try to dominate their weaker neighbors. If democratic governments are not committed to the abolition of such regimes—sooner or later, by some means or other—then their foreign policies are not worth much. For twenty years following Truman's speech, every American administration had to deal with this basic imperative. This meant that every public policy, no matter how domestic in scope or intent, was obliged to answer to the question of whether it aided or retarded the goal of ridding the world of totalitarianism.

The target of the Truman Doctrine was not, explicitly, Soviet Communism. It was totalitarianism, and in the first two decades of the Cold War, many people—public officials, intellectuals with an influence on policy, leaders of cultural institutions and private foundations—believed that art and ideas were an important battleground in this struggle. By various means, covert as well as transparent, they supported the production and dissemination of Western, usually American, cultural products of virtually all types.[1] In the beginning, the targets of these efforts were Western European nations and Japan; after 1956, largely because of a consensus that the Cold War in Europe had been won, attention shifted to the decolonizing world.

---

1. See Frank A. Ninkovich, *The Diplomacy of Ideas: U.S. Foreign Policy and Cultural Relations, 1938–1950* (Cambridge: Cambridge University Press, 1981); Serge Guilbaut, *How New York Stole the Idea of Modern Art: Abstract Expressionism, Freedom, and the Cold War*, trans. Arthur Goldhammer (Chicago: University of Chicago Press, 1983); Naima Prevots, *Dance for Export: Cultural Diplomacy and the Cold War* (Hanover: Wesleyan University Press, 1998); Frances Stonor Saunders, *The Cultural War: The CIA and the World of Arts and Letters* (New York: New Press, 1999); David Caute, *The Dancer Defects: The Struggle for Cultural Supremacy during the Cold War* (Oxford: Oxford University Press, 2003); Volker R. Berghahn, *America and the Intellectual Cold Wars in Europe: Shepard Stone between Philanthropy, Academy, and Diplomacy* (Princeton, NJ: Princeton University Press 2003); Penny M. Von Eschen, *Satchmo Blows Up the World: Jazz Ambassadors Play the Cold War* (Cambridge, MA: Harvard University Press, 2004); Michael L. Krenn, *Fall-out Shelters for the Human Spirit: American Art and the Cold War* (Chapel Hill: University of North Carolina Press, 2005); *Be-Bomb: The Transatlantic War of Images and All That Jazz, 1946–1956* (Barcelona: Museo d'Art Contemporani de Barcelona, 2007); Hugh Wilford, *The Mighty Wurlitzer: How the CIA Played America* (Cambridge, MA: Harvard University Press, 2008); John B. Hench, *Books as Weapons: Propaganda, Publishing, and the Battle for Global Markets* (Ithaca, NY: Cornell University Press, 2010); Greg Barnhisel, *Cold War Modernists: Art, Literature, and American Cultural Diplomacy* (New York: Columbia University Press, 2015).

Because of the superior geopolitical leverage enjoyed by the United States after the war, and because of the economic weakness of Western Europe and Japan and the isolation of Eastern Europe, American involvement sped the development, already underway before the war in parts of Europe, of American-style economies of mass production and consumerism. And American-made cultural products, particularly American entertainment, came to dominate European markets.[2]

This is only half the story, though. The same social forces that were producing changes in American life were producing parallel changes everywhere else. They did not stop or start at any border. Down on the ground, relations between American and non-American art and thought were various, nuanced—above all, dialectical.[3] Japanese and European artists brought

---

2. See Tony Judt, *Postwar: A History of Europe since 1945* (New York: Penguin, 2006), 350–53; Victoria de Grazia, *Irresistible Empire: America's Advance through Twentieth-Century Europe* (Cambridge, MA: Harvard University Press, 2005); William Hitchcock, "The Marshall Plan and the Creation of the West," in *The Cambridge History of the Cold War*, vol. 1: *Origins*, ed. Melvyn P. Leffler and Odd Arne Westad (Cambridge: Cambridge University Press, 2010), 172–73; Volker Berghahn, *The Americanization of West German Industry, 1945–1973* (New York: Berg, 1986); Brian Angus McKenzie, *Remaking France: Americanization, Public Diplomacy, and the Marshall Plan* (New York: Berghahn Books, 2005).

3. See Ralph Willett, *The Americanization of Germany, 1945–1949* (London: Routledge, 1989); Irwin M. Wall, *The United States and the Making of Postwar France, 1945–1954* (Cambridge: Cambridge University Press, 1991), 96–126; Richard F. Kuisel, *Seducing the French: The Dilemma of Americanization* (Berkeley: University of California Press, 1993); R. Kroes, R. W. Rydell, and D. F. J. Bosscher, eds., *Cultural Transmissions and Receptions: American Mass Culture in Europe* (Amsterdam: VU University Press, 1993); Reinhold Wagnleitner, *Coca-Colonization and the Cold War: The Cultural Mission of the United States in Austria after the Second World War* (Chapel Hill: University of North Carolina Press, 1994), 275–96; Richard Pells, *Not Like Us: How Europeans Have Loved, Hated, and Transformed American Culture since World War II* (New York: Basic Books, 1997), 37–262; Uta G. Poiger, *Jazz, Rock, and Rebels: Cold War Politics and American Culture in a Divided Germany* (Berkeley: University of California Press, 2000); Heide Fehrenbach and Uta G. Poiger, eds., *Transactions, Transgressions, Transformations: American Culture in Western Europe and Japan* (New York: Berghahn Books, 2000); Reinhold Wagnleitner and Elaine Tyler May, eds., *"Here, There, and Everywhere": The Foreign Politics of American Popular Culture* (Hanover: University Press of New England, 2000), 83–216; Jessica C. E. Gienow-Hecht, "Shame on US? Academics, Cultural Transfer, and the Cold War—A Critical Review," *Diplomatic History* 24 (2000): 465–94; Alexander Stephan, ed., *The*

the avant-garde to the United States; European thinkers explained American democracy to Americans; anti-colonial leaders in South Asia, the Caribbean, and Africa taught American Blacks how to think about racial discrimination and oppression. American culture did not gain stature after 1945 because the world became Americanized. It gained stature because the world remade America.

Cultural history presents its own periodization problems. When we are trying to understand a work of art or a philosophical movement, we think of it as belonging in a chain with other works or movements, and that chain can run backward for decades. How do you limit the scope of the backstory? After I finished writing *The Free World*, I ended up cutting tens of thousands of words of backstory.

There is also the problem that the arts develop (if that is the right word) at different rates. Changes in the art world (painting and sculpture) between 1945 and 1965 were dramatic; changes in literary fiction and the publishing world were not. Those changes were just starting around 1965. Changes in Hollywood movies (though not in European cinema) begin even later. We can describe the trajectories of specific artistic media, but it's hard to generalize about culture as a whole.

In political history, interpretations of the Cold War have gone through three phases. In the beginning, American historians explained the Cold War as a product of Soviet actions, and specifically of the behavior of Josef Stalin. Starting in 1959, with the publication of William Appleman Williams's *The Tragedy of American Diplomacy*, Cold War history entered a revisionist phase. The Cold War was explained as a product of deliberate American policy, and specifically as an instrument of American business. The revisionist view was succeeded by what is called the post-revisionist view, beginning

---

Americanization of Europe: Culture, Diplomacy, and Anti-Americanism after 1945 (New York: Berghahn Books, 2006); Jessica C. E. Gienow-Hecht, "Culture and the Cold War in Europe," in *Cambridge History of the Cold War*, vol. 1, 398–419; Howard L. Malchow, *Special Relations: The Americanization of Britain?* (Stanford, CA: Stanford University Press, 2011).

with John Lewis Gaddis's *The United States and the Origins of the Cold War*, in 1972. Post-revisionism sees the Cold War as the consequence of actions on both sides. Although this interpretation appears to allow for the possibility that things might have turned out differently, in effect, post-revisionism tends to see the Cold War as inevitable.

I think that cultural history was stuck for a long time in a revisionist phase. To generalize, a little unfairly, cultural history was affected by an anti–Cold War politics that regarded the promotion of a neoliberal, pro-business ideology as more dangerous than the prevention of the emergence of totalitarian regimes in the rest of the world. Like, probably, most historians, I think the United States militarized the conflict with the Soviet Union unnecessarily, that it exaggerated the existential threat, and that it allowed a paranoid anti-Communist rhetoric to suffuse public life. But the policy of containment, as originally conceived, was not mistaken. It just turned out to be inadequate in Southeast Asia.

The United States also, of course, interfered with the internal politics of other countries and engaged in covert funding of ostensibly nongovernmental organizations, compromising their members and activities. (This, too, was a policy extrapolated from the theory of containment.) But "covert" does not mean underhanded. The CIA is part of the executive branch. Funding covertly through the agency enabled the government to support artists and organizations that taxpaying voters might have disapproved of. The failure of the *Advancing American Art* exhibition in 1946–1947, mounted by the State Department, convinced officials that the government had to funnel support for some kinds of cultural diplomacy through other conduits. So it did.

Because any government effort, covert or official, has tended to be regarded by some cultural historians as hegemonic, some tendentiousness can creep into Cold War histories. Frances Stonor Saunders's *The Cultural Cold War* was published in 1999 and is frequently cited. Her book is chiefly about the role the CIA played in cultural diplomacy. I read the book when it came out and was excited by it. I expected when I started writing my own book that I would duplicate many of Saunders's findings. Instead, I found her book to be surprisingly unreliable.

For example, here is a passage from the book on George Kennan:

> In a speech to the National War College in December 1947, it was Kennan who introduced the concept of "the necessary lie" as a vital constituent of American post-war diplomacy. The Communists, he said, had won a "strong position in Europe, so immensely superior to our own . . . through unabashed and skillful use of lies. They have fought us with unreality, with irrationalism. Can we combat this unreality successfully with rationalism, with truth, with honest, well-meant economic assistance?" he asked. No. America needed to embrace a new era of covert warfare to advance her democratic objective against Soviet deceit.[4]

Saunders provides the following citation for the quotation from Kennan: George Kennan, National War College Address, December 1947, quoted in *International Herald Tribune*, 28 May 1997.[5]

There is no quotation from Kennan in the *International Herald Tribune* for May 28, 1997, or in any other issue of that paper. And although Kennan did speak at the National War College in December 1947, he did not say the words she quoted. He *did* say this, in a talk at the National War College on June 18, 1947:

> [The Communists'] strong position in Europe, so immensely superior to our own . . . not by economic means . . . [but] through unabashed and skillful use of lies. They have fought us with unreality, with irrationalism. Can we combat this unreality successfully with rationalism, with truth, with honest, well-meant economic assistance? Perhaps not. But these are

---

4. Frances Stonor Saunders, *The Cultural Cold War: The CIA and the World of Art and Letters* (New York: New Press, 1999), 38–39.
5. Saunders, *The Cultural Cold War*, 433n13.

the only weapons we possess, short of war. We hope that at least these weapons will serve to strengthen the resistance of other people to the lure of unreality.[6]

This is the opposite of what Saunders said Kennan said. Kennan never used the words "necessary lie" in his talks to the National War College. So far as I can discover, he never used the phrase anywhere.

There are a number of similar misinterpretations and misattributions in *The Cultural Cold War*. Some have to do with what has become, since the early 1970s, the central case for revisionism, abstract expressionism.[7] The argument is that the CIA and the Museum of Modern Art (MoMA) weaponized abstract expressionism in a propaganda war. But a lot of this argument depends on guilt by association (that is, ties between museum officials and the intelligence community, which certainly existed) and faulty evidence. Saunders claims, for example, that Alfred Barr, the founding director of the MoMA, referred to abstract expressionism as "benevolent propaganda for foreign intelligentsia." That phrase was actually used by a *critic* of the museum's policies, Max Kozloff, in an article published in 1973.[8] Kozloff was a revisionist.

It is not the case that abstract expressionism swamped Europe in the 1950s. In 1958, there was not a single painting by Willem de Kooning, Franz Kline, Robert Motherwell, or Mark Rothko in a European museum. No European museum purchased a Pollock until 1961. Mark Rothko did not have a solo exhibition in Europe until 1961; Willem de Kooning did not have one

6. George F. Kennan, *Measures Short of War: The George F. Kennan Lectures at the National War College, 1946–47*, ed. Giles D, Harlow and George C. Maerz (Washington, DC: National Defense University Press, 1991), 212.
7. See Max Kozloff, "American Painting during the Cold War," *Artforum*, May 1973, 43–54; Eva Cockroft, "Abstract Expressionism, Weapon of the Cold War," *Artforum*, June 1974, 39–41; David and Cecile Shapiro, "Abstact Expressionism: The Politics of Apolitical Painting," *Prospects* 3 (1977): 175–214; Guilbaut, *How New York Stole the Idea of Modern Art*, 168–74; Saunders, *The Cultural Cold War*, 255–78.
8. Kozloff, "American Painting," 44.

until 1967. By then, Andy Warhol had had ten European exhibitions and Robert Rauschenberg had had fifteen. The American art that "conquered the world" was Pop Art, not abstract expressionism. The music was rock 'n' roll, not jazz. The literature was Beat literature, not modernist poetry. The American government was not involved with any of those agents of change.[9]

We may get some perspective by comparing American culture in the early Cold War period to British culture in the nineteenth century. In nineteenth-century Britain, you would expect to find evidence of the fact that Britain was an imperial state everywhere you looked—in art, in poetry, in philosophy, in the culture of everyday life. You can't subtract the British empire from British culture, and, similarly, you can't subtract the Cold War from postwar American culture. The historical problem is to explain the difference it made. How did the Cold War shape art and ideas in the postwar period, and how was it a factor in changing those things?

My thesis is that the questions artists and writers raised (What is a painting? What is a poem?) seemed urgent, and the answers mattered, for Cold War reasons. The government's message was that in the Free World, unlike under Communist rule, the state does not tell you what to write or how to paint. But the effectiveness of this message obviously depended on the quality of cultural goods. If the paintings and music produced by a "free society" were inferior, immature, somehow "not art," then they were a bad advertisement for liberal democracy.

---

9. See Michael Kimmelman, "Revisiting the Revisionists: The Modern, Its Critics, and the Cold War," *Studies in Modern Art* 4 (1994): 39–55; Robert Burstow, "The Limits of Modernist Art as a 'Weapon in the Cold War': Reassessing the Unknown Patron of the Monument to the Unknown Political Prisoner," *Oxford Art Journal* 20 (1997): 68–80; David Caute, *The Dancer Defects: The Struggle for Cultural Supremacy during the Cold War* (Oxford: Oxford University Press, 2003), 539–67; Irving Sandler, "Abstract Expressionism and the Cold War," *Art in America*, June/July 2008, 65–74; Kathryn Anne Boyer, "Political Promotion and Institutional Patronage: How New York Displaced Paris as the Center of Contemporary Art, ca. 1935–1968" (PhD diss., University of Kansas, 1994); Hiroko Ikegami, *The Great Migrator: Robert Rauschenberg and the Global Rise of American Art* (Cambridge, MA: MIT Press, 2010); and Catherine Dossin, *The Rise and Fall of American Art, 1940s–1980s: The Geopolitics of Western Art Worlds* (Burlington, VT: Ashgate, 2015).

In this respect, the Cold War charged the cultural atmosphere. It raised the stakes. The more artistic and intellectual expression became unfettered, by changes in style, in the audience, and in the legal environment, the more important it was that the art be genuine art. On this dimension, the aestheticism of the period did have a politics. Who was thinking through those politics at any given time is another question. I doubt that Jackson Pollock asked himself whether his paintings were good advertisements for the American way of life.

When you subtitle a book *Art and Thought in the Cold War*, you are singling out one variable—the geopolitical situation—in what should be a multivariable analysis. I found that other factors were generally more determinative of cultural developments than superpower relations. The crucial event was the rise of Hitler. That set in motion a period of European out-migration that lasted until the United States entered the war, in 1941. Some of those emigrants ended up in the United Kingdom and the British Commonwealth. Many came to the United States and had a significant impact on intellectual life and arts practice. Almost none of them would have emigrated if Hitler had not become chancellor in 1933. Most of them had no prior interest in or much respect for the United States as a civilization. They did admire the official policy of freedom of expression, and they benefited from it.

A second powerful social movement was decolonization. That is, really, the big story: between 1945 and 1970, most former European colonies became independent sovereign states governed by nonwhite people. Decolonization redrew the international map. It changed the way social scientists conceived of human difference. It also put pressure on the American government to redress racial injustice in the United States.[10]

---

10. See Mary L. Dudziak, *Cold War Civil Rights: Race and the Image of American Democracy* (Princeton, NJ: Princeton University Press, 2000). See also John David Skrentny, "The Effect of the Cold War on African American Civil Rights: America and the World Audience, 1945–1968," *Theory and Society* 27 (1998): 237–85; Thomas Borstelmann, *The Cold War and the Color Line: American Race Relations in the Global Arena* (Cambridge, MA: Harvard University Press, 2001); Brenda Gayle Plummer, *Rising Wind: Black*

Migration, decolonization, demographic change, and economic conditions are all long-term factors in cultural history. From the point of view of the individual actor, they are intangibles. Less intangible are changes in the culture industries, changes caused by legal, financial, and technological developments. Between 1945 and 1965, the American culture industries grew dramatically, building infrastructure that enabled the production and dissemination of new kinds of cultural goods, both fine art and popular entertainment: the music industry, the art world, book publishing, the magazine business, museums, universities. These had more to do with the rate and direction of cultural change than the Cold War did.

Was there a Zeitgeist? I'm not sure I believe in the Zeitgeist. What I do believe in is individuals trying to make something—a poem, a song, a painting. When we get inside that process, we see that the ingredients are individual talent, opportunity, a fluctuating conceptual framework, intention, and accident. When you're trying to make sense of an idea or a creative work, you can't ignore any of these elements.

In writing history: everything is potentially relevant, from the ownership of the means of production down to the color the artist painted their toenails. You try to see it all. And the method is the method of the hermeneutic circle. Each fresh detail alters, by a tiny increment, the big picture; the now-altered big picture affects the understanding of every detail. You go back and forth until you think you have got it right. And you historicize—always reminding yourself that you are, as a subjectivity, a product of the history that you are trying to write.

Americans and U. S. Foreign Affairs (Chapel Hill: University of North Carolina Press, 1996), 167–297; Plummer, *In Search of Power: African Americans in the Era of Decolonization, 1956–1974* (Cambridge: Cambridge University Press, 2013); and Paul Gordon Lauren, "Seen from the Outside: The International Perspective on America's Dilemma," in *Window on Freedom: Race, Civil Rights, and Foreign Affairs, 1945–1988*, ed. Brenda Gayle Plummer (Chapel Hill: University of North Carolina Press, 2003), 21–43.

Tales from the Hot Cold War

MARTHA BAYLES

Abstract

Unlike most of the world's film industries, the American system known as Hollywood has never been owned or operated by the national government. Instead, it has been privately owned from the earliest days. This does not mean, however, that Hollywood has been unaffected by the concerns and priorities of Washington, DC. Despite being twenty-seven hundred miles apart, the two cities have cooperated, and contended, on many occasions—especially during times of war. This essay explores the history of that relationship from the 1910s to the present.

Keywords: War War II films, Cold War films, Korean War films, Vietnam War films, Washington and Hollywood

> What truth soldiers would speak/None would hear, and none repeat.
> —Howard Lachtman

Despite being the world's most influential storyteller, Hollywood has rarely been good at dramatizing war. This essay examines this shortcoming through the lens of the "Washington-Hollywood Pact", my term for the volatile but long-lasting partnership between America's government and its dream factory. The intention is to show how, from the earliest days of commercial filmmaking, changes in the American experience of war have shaped the way war has been dramatized by Hollywood—and to argue, through a consideration of specific films, that these changes have so altered the terms of the pact that it no longer represents the best interests of either partner.

https://doi.org/10.3998/gs.3428

The Origins of the Pact

The Washington-Hollywood Pact dates back to 1915, when a landmark decision by the US Supreme Court, *Mutual Film Corporation v. Industrial Commission of Ohio*, defined film as "a business, pure and simple." In effect, this decision denied First Amendment protection to the new medium, thereby exposing it to government censorship at the local, state, and federal levels. As a result, when the United States was preparing to enter World War I in 1917, the studios readily complied with president Woodrow Wilson's request that they produce propaganda films in support of an unpopular war.

The word *propaganda* is not an exaggeration. The films that resulted were unsubtle in the extreme. Indeed, some took their cues from the British government's efforts to stoke war fever by racializing Germans as "Huns" and fabricating reports of German atrocities in Belgium.[1] Newly relocated to Los Angeles from New York, the fledgling film industry was not yet called Hollywood. But by producing crude films like *Escaping the Hun* and *The Kaiser, the Beast of Berlin*, the studios committed themselves to the pact.

At war's end, Washington rewarded Hollywood by permitting the studios to use aggressive tactics, like cartels and block booking, that were illegal at home but useful in cornering overseas markets. These tactics were not the only reason why America was able to replace France as the world's leading film exporter by 1925. Hollywood movies were created to please a highly diverse domestic audience, so they traveled well (and still do). But without the pact, they might not have traveled so far so fast.

World War II bolstered the pact. Focused as usual on the bottom line, the studios sold movies to Fascist Italy and Nazi Germany right through

---

1. See Philip M. Taylor, *Munitions of the Mind*, 3rd ed. (Manchester: Manchester University Press, 2003), 196–97.

1939, ceasing only when Mussolini and Hitler banned all such imports. But as of December 1941, when imperial Japan bombed Pearl Harbor and Germany declared war on America, Hollywood stood ready to produce whatever Washington needed, from Frank Capra's army training films *Why We Fight*, made with an assist from Walt Disney, to hundreds of theatrical features supporting the war aims of America and its allies in more or less artistically successful ways.

With victory came unprecedented power for both partners. Through the Marshall Plan and other aid programs, Washington provided vital assistance to a number of war-torn countries, both allies and former foes, while attaching a proviso requiring the recipients to admit floods of Hollywood films. The films were donated by the Motion Picture Association of America (MPAA) in the hope of prying open new markets. And the pressure to accept them was applied by the Departments of State and Defense in the hope that American-style entertainment would persuade foreign populations that American-style democracy was superior to fascism or communism.

Before war's end in 1944, this policy was set forth in a memo sent to the MPAA by the US Department of State: "In the post-war period, the Department desires to co-operate fully in the protection of American motion pictures abroad. It expects in return that the industry will co-operate wholeheartedly with the government with a view to ensuring that the pictures distributed abroad will reflect credit on the good name and reputation of this country and its institutions."[2] Rarely had the terms of the pact been spelled out so clearly. The tide of war was turning, and Hollywood shared Washington's urgent desire to defeat Hitler. But because victory was not yet assured, the memo's alignment of diplomatic and commercial priorities made sense to both partners.

---

2. Quoted in David Puttnam, *The Undeclared War* (London: Harper Collins, 1997), 212.

Three years later, that alignment made less sense. In 1947, amid rising tensions between the Soviet Union and its former allies America and Britain, the House Un-American Activities Committee (HUAC) began a series of hearings on Communist influence in the movie colony. A decade earlier, a similar committee had investigated Nazi influence, but with little result. This probe was different, because Hollywood had long had its share of Communists and fellow travelers.

In 1947, however, the leftists in Hollywood had reason to assume they were in good standing with Washington. This is because, back in 1941, when president Theodore Roosevelt's Office of War Information (OWI) urged the studios to shore up relations with America's new ally, the Soviet Union, by making pro-Soviet films, it was the leftists who most readily agreed.[3] A salient example would be Howard Koch, a respected screenwriter best known today for his work on the 1942 classic *Casablanca*. In 1941, when the OWI urged Warner Bros. to make a pro-Soviet film based on the memoirs of Joseph E. Davies (who had served as US ambassador to the Soviet Union between 1936 and 1938), Jack Warner, the powerful head of that studio, asked Koch to write the script.

Koch agreed, only to regret it later. Davies's book, *Mission to Moscow*, was astoundingly credulous and naïve. But it was also a bestseller, so in order to secure the rights, Warner Bros. gave Davies complete creative control. The resulting film was a mess—part vanity project by Davies, part Stalinist propaganda. OWI called it "a magnificent contribution to the Government's motion picture program."[4] But the American public stayed away in droves; anti-Communist pundits denounced it as the work of the Kremlin; even the Soviet leadership disliked it. According to the US ambassador, who held a private screening of *Mission to Moscow* for Stalin and his inner circle, Stalin's cronies watched it with "glum curiosity" while the "Leader of Humanity"

---

3. Clayton R. Kopper and Gregory D. Black, *Hollywood Goes to War* (New York: Free Press, 1987), 10.
4. Quoted in Kristin Hunt, "Hollywood Codebreakers: Warner Brothers Embarks on a Disastrous 'Mission to Moscow,'" *Medium*, April 20, 2018.

(a film buff when not ordering mass executions) "was heard to grunt once or twice."[5]

It is therefore ironic that in 1947, when HUAC was holding its hearings, the same Jack Warner who had enlisted Koch to write *Mission to Moscow* gave the committee a list of eighteen "Commies" working in Hollywood that included Koch. When subpoenaed by HUAC, Koch vowed not to remain silent as others had but to defend his cooperation with the war effort. In the end, he was not called to testify, but in 1951 he was blacklisted. He found work in London, but others were less fortunate.[6]

There is some comfort in knowing that in wartime Hollywood, even the most committed Communists found it hard to tell Stalin-sized lies. Today, *Mission to Moscow* rests in the dustbin of history, along with the other pro-Soviet clunkers made at the time. On the scale of injustices committed during the twentieth century, the HUAC probe and the Hollywood blacklist are blips. But that has not stopped the movie colony from making more films about those incidents than about, say, the forced starvation of five million people in Soviet Kazakhstan, Russia, and Ukraine between 1930 and 1933.

The First Hot War: Korea

When the Marshall Plan was launched in 1948, its two main goals, according to historian Thomas H. Guback, were "to strengthen faltering economies against risings from the left" and "to open and maintain markets for American films, . . . seen as propaganda vehicles for strengthening western European minds against pleas from the left."[7] In 1948 the Cold War was just

---

5. William H. Standley, telegram to the US Secretary of State, May 25, 1943. Quoted in Todd Bennett, "Culture, Power, and *Mission to Moscow*: Film and Soviet-American Relations during World War II," *Journal of American History* (September 2001).

6. "Howard Koch," Bard Archives and Special Collections, accessed October 15, 2022, https://www.bard.edu/archives/voices/koch/koch.php.

7. Thomas H. Guback, *The International Film Industry* (Bloomington: Indiana University Press, 1969).

beginning, and by the terms of the pact, Hollywood would soon be playing its part by producing "propaganda vehicles" to support the anti-Communist cause.

Those terms were about to change, however. In June 1950, one hundred thousand North Korean troops crossed the 38th parallel into South Korea. The United Nations (UN) responded by sending a coalition force commanded by US Army general Douglas MacArthur, and the Cold War's first hot war began. This was labeled a "police action" because it did not involve a formal declaration of war. But it was definitely a war: three years of bloody combat, ordeals of survival under unspeakably harsh conditions, dramatic reversals of fortune—all rich fodder for audience-pleasing movies.

Or so it would seem. But in fact, the Korean conflict was slow to inspire Hollywood. The first to address the topic was Samuel Fuller, a World War II veteran turned filmmaker, who in 1951 began to make gritty, low-budget films about Korea. First and most successful was *The Steel Helmet*, about a diverse group of US soldiers who unite against enemy efforts to defeat them not just militarily but psychologically by playing on their racial differences. Produced for $104,000 in a Los Angeles park, with a plywood tank, a mist machine, and twenty-five UCLA students as extras, *The Steel Helmet* grossed over $2 million.[8]

Two years later, the bestselling author James Michener published *The Bridges at Toko-Ri*, a novel drawing an emotional connection between World War II and Korean conflict. The main character, Lt. Harry Brubaker (William Holden), is a navy fighter-bomber pilot who, despite having flown many missions in the Pacific during World War II, has been called back to serve in Korea. He resents this, because as he tells an admiral who is about to give him a particularly dangerous mission, "I was one of the few, Admiral, at New Guinea, Leyte, Okinawa. Why does it have to be me again?"[9]

---

8. Samuel Fuller, *A Third Face* (New York: Alfred A. Knopf, 2002), 257–58.
9. Mark Robson, *The Bridges at Toko-Ri* (Los Angeles: Paramount Pictures, 1954). See also "The Bridges at Toko-Ri," Scripts, accessed November 24, 2022, https://www.scripts.com/script.php?id=the_bridges_at_toko-ri_19853&p=2.

Brubaker then expresses what many Americans, not just veterans, felt about Korea: "Militarily, this war is a tragedy. I think we ought to pull out." The admiral replies with a pitch-perfect statement of the anti-Communist cause: "That's rubbish, son, and you know it. If we did, they'd take Japan, Indochina, the Philippines. Where would you have us make our stand, the Mississippi?" This dialogue, followed by a powerfully realized film about the bombing of heavily fortified bridges in the mountains of North Korea, is precisely what Hollywood was expected to produce under the terms of the pact. But it's important to remember that *The Bridges at Toko-Ri* could not have been made without the cooperation of the US Navy. In other words, Washington had a veto on this one.

The next major film about Korea, the 1959 battle piece *Pork Chop Hill*, is based on the true story of a US Army platoon being decimated by Chinese forces while trying to take control of an outpost marking the dividing line between North and South. In many ways the film harks back to the patriotic war films of the 1940s. But it also strikes an ambivalent note characteristic of the director, Lewis Milestone, who in 1930 had made his name with a screen adaptation of *All Quiet on the Western Front*, Erich Maria Remarque's pacifist novel about World War I.

In 1945, Milestone had also directed *A Walk in the Sun*, a portrait of US soldiers breaking down under pressure during the Allied invasion of Italy. To make such a film was a bold step, because while such "burnouts," as they were called, were common in World War II, they rarely showed up in Hollywood films. According to General Harold K. Johnson, a survivor of the Bataan Death March and a Japanese POW camp, the fact that troops were in the war "for the duration . . . resulted in an enormous number of battlefield fatigue cases," as soldiers "just break down under that kind of unremitting pressure."[10]

---

10. Mark DePue, "Vietnam War: The Individual Rotation Policy," *Vietnam Magazine*, December 2006.

Milestone was neither called before HUAC nor blacklisted. But he was "grey-listed," or denied work because of his suspected Communist sympathies, which may help to explain his ambivalence in *Pork Chop Hill*. The film ends with a voice-over tribute to "those who fought there . . . millions live in freedom today because of what they did." This echoes the casus belli—the "why we fight"—portrayed in hundreds of Hollywood World War II films: *We fight to defend democracy against Fascism*. But for Hollywood in the 1950s, the fight against *Communism* did not have the same resonance.

Further, it was clear in 1959 that the Korean conflict had been neither won nor lost, but rather frozen in place as part of the Cold War. What Hollywood needed now was a casus belli better suited to the times. *Pork Chop Hill* supplied it, in a comment by platoon commander Lt. Joe Clemons (Gregory Peck) to a fellow officer who asks if taking the hill had been worth it: "Worth what? It hasn't much military value. I doubt if any American'd give you a dollar for it. Probably no Chinese'd give you two bits." But then Clemons adds, "The value's changed somehow, sometime. Maybe when the first man died."[11]

Unlike the voice-over, which affirms that *they fought to defend democracy against Communism*, this comment suggests a new casus belli: *They fought to justify and avenge the deaths of their comrades*. This is hardly a new idea. In Homer's *Iliad*, Achilles refuses to fight after being insulted by the Greek leader Agamemnon, returning only when the Trojans kill his beloved friend Patroclus. In the heat of battle, soldiers do not think about long-range objectives; they think about their friends. And when they display courage, it is more out of loyalty to their friends, or fear of letting them down, than commitment to any abstract ideal. Sociologists call this "unit cohesion," and no army can succeed without it. What is new in *Pork Chop Hill* is the notion that unit cohesion by itself might be a sufficient casus belli.

---

11. Lewis Milestone, *Pork Chop Hill* (Beverly Hills, CA: Melville Productions, 1959). See also "Pork Chop Hill (1959)," Movie Scripts, accessed November 24, 2022, https://stockq.org/moviescript/P/pork-chop-hill.php.

Solidarity among comrades is a vital ingredient in every World War II film made in Hollywood, from the earliest propaganda to recent gems like *Saving Private Ryan* (1998) and *Band of Brothers* (2001). In the context of that "good war," it works well to foreground the tribulations of a small group of soldiers, because in the background there is always the struggle to defeat Hitler. Indeed, *Band of Brothers* includes an episode called "Why We Fight," in which the burned-out, hardened soldiers in Easy Company find new purpose after stumbling upon a Nazi death camp. The scene is powerful but feels a bit didactic, because we already know why they fight.

This World War II formula of "comrades in the foreground, worthy cause in the background" also served to promote democracy by portraying American soldiers as ordinary citizens from diverse backgrounds (more diverse over time), who embody democratic virtues. The soldiers are not all heroic, but the ones we admire tend to fit a certain profile. They show respect, not slavish obedience, to their superior officers; they think and speak for themselves; they know when to follow orders and when to seize the initiative; and they do not hesitate to make the ultimate sacrifice when the cause of freedom requires it.

The Second Hot War: Vietnam

Between 1955, when America stepped in to support the South Vietnamese government after the defeat of the French colonial forces at Dien Bien Phu, and 1975, when Saigon fell to the combined forces of the Communist North, Hollywood struggled to find a compelling casus belli for America's involvement in Vietnam. As a result, most of the relevant films were made after 1975, and the majority reflect the discontent, futility, and profound alienation that had come to be felt by all but its most ardent supporters.

It was different in the early 1960s, when young Americans marched off to Vietnam with visions of John Wayne dancing in their heads. This sounds like a cliché, but according to the distinguished British military historian

Richard Holmes, it was true. After citing numerous sources on this point, Holmes concludes that "middle-ranking infantry officers in Vietnam in the late 1960s would have been in their early teens when *The Sands of Iwo Jima* first appeared; it is, perhaps, not surprising that its impact was so tremendous."[12] Thus, the first Hollywood film made about Vietnam was a John Wayne vehicle called *The Green Berets*.

Produced in the style of 1949, set in 1963 during the confident early phase of American involvement, and released into the havoc of 1968, *The Green Berets* landed with a resounding thud. In 1949 audiences had cheered at *The Sands of Iwo Jima*, especially the ending, which reenacts the famous photograph of US Marines raising the stars and stripes on Mount Suribachi (itself a reenactment, no photographer having been present when the event actually occurred). But in 1968 audiences jeered at *The Green Berets*, both for its general klutziness and for filling the screen with incongruities such as the stealthy guerilla fighters of the Viet Cong marching in close formation, and the sun setting in the east.

More attuned to the mood of 1968 was Robert Altman's *M*A*S*H* (1970), about three army surgeons who, when not up to their elbows in blood, amuse themselves by creating chaos for everyone around them. The film's ostensible setting is a field hospital in Korea, but its popularity rested on the assumption that it was really about Vietnam. In 1970, when *M*A*S*H* was released (or, as Altman once quipped, when it "escaped"), the production studio, Twentieth Century Fox, was heavily invested in two high-prestige films, *Patton* and *Tora! Tora! Tora!*, about World War II. There were no films about Vietnam on the horizon, so when funky, low-budget *M*A*S*H* grossed millions more at the box office, Hollywood sat up and took notice.[13]

---

12. Richard Holmes, *Firing Line* (London: Pimlico, 1994).

13. *M*A*S*H* grossed $81.6 million at the domestic box office, compared to *Patton* ($62.5 million) and *Tora! Tora! Tora!* ($29.5 million). The Numbers, "Weekend Predictions: Halloween Ends Eyes $50-Million Opening," Nash Information Services, October 14, 2022, https://www.the-numbers.com/movie/.

The late 1960s and early 1970s were the best of times for Hollywood, in the sense of finally gaining artistic freedom under the First Amendment. In 1968, after a series of Supreme Court decisions overturned the 1915 definition of film as "a business, pure and simple," the MPAA dismantled the Production Code Authority (PCA), also known as the Hays Office, which the studios had created in the 1930s to ward off government censorship by a prerelease content-monitoring process. In the PCA's stead, the MPAA set up the age-graded ratings system still in place today.

This new freedom nurtured countless brilliant films that could not have been made under the Production Code. But it also fostered an overreliance on explicit sex and graphic violence to enliven material that might otherwise hold scant interest. In the case of war films, more freedom meant rapid progress in the simulation of mayhem. No longer did soldiers topple gracefully when struck by imaginary bullets; the blood squib made it possible to show arterial spurting with every hit. No longer did exploding artillery shells toss rag-doll bodies into the air; latex and dye made it possible to show bursting heads, severed limbs, and random body parts raining down on both the quick and the dead.

One of the first Vietnam-related films, Martin Scorsese's celebrated *Taxi Driver*, was released in 1975, when the troops had barely returned home. A fever dream of loneliness and paranoia in the filthy streets of mid-1970s Manhattan, the film stars Robert De Niro as Travis Bickle, a troubled loner suffering from insomnia, who passes his nights driving a taxi, visiting porn shops, and raging at the exploitation of teenaged prostitutes. Today this mental state is called post-traumatic stress syndrome (PTSD), and it was all too common among Vietnam veterans, for reasons discussed below.

But instead of a sympathetic portrait, the screenwriter Paul Schrader made Travis a bloodthirsty psychopath. Why did Schrader do that? Possibly he was inspired by *Targets*, Peter Bogdanovich's directorial debut. Released in 1968, the same year as *The Green Berets*, *Targets* is loosely based on the story of mass shooter Charles Whitman, who in 1966 killed fourteen people on the University of Texas's Austin campus. Whitman was clearly suffering from mental illness. But for good measure, and perhaps as a riposte to John Wayne, Bogdanovich made his fictionalized shooter a Vietnam vet.

Around the same time, Schrader wrote the script for another film, *Rolling Thunder* (1977), which likewise combines state-of-the-art gore with a lead character who after being traumatized in Vietnam becomes a ticking time bomb. About these choices, we need not wonder. In the late 1970s, a blood-drenched film about a mere killer or psychopath might have been scary and depressing, but it would not have set the world thrumming with speculation about the filmmaker's intentions and debate about the deeper political significance of every frame. Make that character an American who fought in Vietnam, and the thrumming would go on for decades.[14]

Gimme (Tax) Shelter

Earlier I suggested that the late 1960s and early 1970s were the best of times for Hollywood as it achieved a long-sought degree of artistic freedom. But those years were also the worst of times, because the major studios were hemorrhaging money on lavish musicals and other old-fashioned movies that, to say the least, failed to resonate with domestic audiences deeply divided over the Vietnam war, campus protests, urban riots, and political assassinations. The studios badly needed bailouts from Washington. But given their lack of enthusiasm for the government's anti-Communist cause, was it reasonable of them to expect such help?

Surprisingly, it was. This is because Washington and Hollywood are like an old married couple who quarrel at home but are deeply united in their outward-facing dealings with the world. As the war in Vietnam escalated, and Hollywood stopped making films like *The Green Berets* and began

---

14. In passing, I should mention that *The Ninth Configuration* (1979), written and directed by William Peter Blatty, also features a traumatized vet. But as part of a trilogy exploring questions of good and evil, faith and nihilism, the film contains very little violence and has no political agenda to speak of.

making the kind that portray US soldiers as dangerous killers, Washington did not flinch. On the contrary, it behaved like a loyal spouse who will do anything to save a bad marriage.

For example, in 1969, Congress, with the support of the Nixon administration, began to offer tax credits to the studios, enabling them to set up lucrative profit shelters for wealthy investors. Two years later, the 1971 Revenue Act allowed the studios to deduct 7 percent (subsequently 10 percent) of production investment from their overall corporate tax (up to a 50 percent limit) and to carry that forward for seven years. Further sweetening the deal was the outcome of a lawsuit brought by Disney that made the measure retroactive to 1962.[15] As Alan Hirschfield, then president of Columbia Pictures, subsequently testified before Congress, "The availability of this kind of financing is the single most important occurrence in the recent history of the industry."[16]

This Washington help was also "not without a touch of irony," writes filmmaker and former Columbia CEO David Puttnam, because "for years, the studios had fulminated against the preferential tax incentives for film production offered by *foreign* governments" (emphasis added). The irony is further compounded by the benefit these legislative favors conferred on the so-called American New Wave in the 1970s. As Puttnam observes, many of the more "bold and adventurous" films of that decade, including *Taxi Driver*, "were financed using tax shelter money."[17]

Coincidence is not cause, but this help from Washington was followed by a series of films pushing back against the negative stereotype of Vietnam vets as ticking time bombs. In *Coming Home* (1978), a marine wife (Jane Fonda) falls in love with a paraplegic vet (Jon Voigt) while her marine captain husband (Bruce Dern) is serving in Vietnam. A love triangle set in sunny California, the film expresses considerable sympathy for wounded and traumatized

---

15. David Puttnam, *The Undeclared War* (New York: Harper Collins, 1997), 267.
16. Quoted in David A. Cook, *Lost Illusions*, vol. 9: *History of the American Cinema*, ed. Charles Harpole (New York: Scribner's, 2000), 12–13.
17. Puttnam, *The Undeclared War*, 268–69.

veterans—as long as they share Fonda's and her fellow filmmakers' view of the war as criminal aggression against a peace-loving Asian nation.

More subtle is Michael Cimino's *The Deer Hunter*, which won the 1978 Oscar for Best Picture, bestowed upon the director and his team by none other than John Wayne, in his last public appearance. It is a remarkable film, starring Robert De Niro as Michael, one of three blue-collar grunts from western Pennsylvania who are captured by Viet Cong and forced to play Russian roulette for their captors' sadistic pleasure. Michael's friends break down, but he holds out long enough to engineer an escape. Michael's courage and loyalty are a welcome change from the deranged stereotype. But for all its virtues, *The Deer Hunter* evades the question of whether unit cohesion can survive without a compelling casus belli. The main characters were buddies long before Vietnam, and while the other two can clearly count on Michael, there are no scenes of combat to test whether Michael can count on *them*.

In 1979 a very different film appeared, *Apocalypse Now*, directed by Francis Ford Coppola and purportedly based on *Heart of Darkness*, Joseph Conrad's 1899 novella about the degradation of human nature in the Belgian Congo, which under the rule of King Leopold II was the most nightmarish colony in Africa. Without launching into a proper comparison, I will say that the film neglects the deeper meaning of the novella, which is that absolute power destroys both the powerful and the weak, and instead focuses on the surface plot of a civilized European going rogue among "savages." Needless to say, this plot is an awkward fit with the American war in Vietnam, which even at its nadir did not involve any jungle despotisms in which white men were worshiped as gods by the "natives."

Dereliction of Duty

Apocalypse Now widened the gap between the facile anti-war sentiment in Hollywood and the complex emotions of millions of Americans who had supported the war and seen their loved ones bear the brunt of its savagery, only to be scarred by defeat and disrespect. For that audience, a more relevant film was *Go Tell the Spartans* (1978), based on a fictionalized account of a disastrous US Special Forces mission to the Central Highlands in 1964. In the words of a reporter who accompanied that mission, the unit leaders were "admirably trained and motivated," but the orders they were given showed zero forethought and caused the deaths of too many good men.[18]

Go Tell the Spartans did respectably at the box office, but its message—that the real problem in Vietnam was dereliction of duty higher up in the chain of command—got lost in the noise of a rapidly polarizing America.[19] In 1982, that message exploded onto the screen in the hugely popular character of Rambo, a maverick super-warrior whose main battles are with US military and law enforcement. In *Rambo: First Blood*, the first of five films starring Sylvester Stallone, the allure of big weapons and loud explosions is reinforced by a darker, more compelling theme: the resentment of veterans toward Washington for having failed (in Rambo's words) to "let us win."

Actually, Rambo says "let *me* win," which speaks volumes about the legacy of Vietnam. Rambo is a solo act, more comic-book hero than real-life soldier, and his extreme individualism is not a remedy for what went wrong in Vietnam—it is a symptom. From the veterans' perspective, one of the worst aspects of service in Vietnam was the policy of "individual rotation," which instead of training and supporting small and cohesive combat units moved individuals in and out of the field in a manner that seemed deliberately designed to prevent much-needed friendship and trust.

---

18. Daniel Ford, *The Only War We've Got: Early Days in South Vietnam* (Durham, NH: Warbird Books, 2012), https://www.warbirdforum.com/onlywar.htm.

19. *Dereliction of Duty* is of course the title of US Army lieutenant general H. R. McMaster's 1997 book blaming America's defeat on two successive presidents, John F. Kennedy and Lyndon B. Johnson, and their top military and civilian advisors, who led the country into a conflagration that lasted twenty years, cost $139 billion (by official estimate; the true figure is doubtless much larger), and resulted in the deaths of fifty-eight thousand American soldiers, two hundred and fifty thousand South Vietnamese soldiers, five thousand soldiers from US-allied nations, 1.1 million North Vietnamese and Viet Cong fighters, and two million civilians.

The reasons for this policy were many. President Lyndon Johnson did not want to call up the US Reserves and the National Guard, because that would have required Congress to debate the war instead of his social-welfare agenda. The junior officers, focused as they were on the Cold War in Europe, needed heat-of-battle experience to further their careers. The Selective Service legislation limited the tour of duty required of a draftee to two years, which, because of the time spent in training, effectively meant one year. And the senior command, mindful of the burnout issue in World War II, hoped to improve morale by letting every soldier know the end date of his tour. Taken separately, these reasons made sense. But taken together, they led to disaster. As stated in *Vietnam Magazine*, "The quintessentially American emphasis on the individual had replaced the soldier's ethos of selfless service, an ethos that called for soldiers to subordinate their own selfish interests to the welfare of the group."[20]

The Cold War ended in 1989, and it was during the preceding few years that Hollywood came closest to getting Vietnam right. The first film to dramatize the destructive effects of the individual rotation policy was Oliver Stone's *Platoon* (1986), praised by vets for its intense evocation not only of combat but of the discomfort caused by everything from monsoons to mosquitoes. The story begins with the arrival of Chris (Charlie Sheen), a fresh-faced "cherry" lieutenant whose inexperience puts him at the mercy of two sergeants: cruel, racist, scar-faced Barnes (Tom Berenger) and kindly, wise, graceful Elias (Willem Defoe).

Stone's films are not known for their moral complexity, so Barnes gets all the nasty jobs, such as interrogating terrified villagers, while Elias gets all the nice ones, such as tracking North Vietnamese Army (NVA) regulars through the sun-dappled greenwood. The soldiers also divide along the same tidy lines, with the bigoted whites sharing Barnes's taste for booze and killing and the soulful blacks smoking herb with Elias and questioning why they are there. Overall, *Platoon* is a gripping film that deserves credit for accurately

---

20. DePue, "Vietnam War."

portraying the problem of rotating command. But it is also propaganda in the sense of blurring the distinction between soldierly discontent with the way the war is being fought and political opposition to its larger purpose.

A similar sleight of hand occurs in *Full Metal Jacket*, Stanley Kubrick's 1987 film about a platoon of US Marines slogging their way through Da Nang and Hue during the 1968 Tet Offensive. Based largely on *The Short-Timers*, a profane, hard-bitten, terrifying novella by war correspondent Gustav Hasford, the film centers on the relationship between two US Marines: a reporter-turned-soldier nicknamed "Joker" (Matthew Modine) and his levelheaded Texan friend "Cowboy" (Arliss Howard). Having bonded in boot camp against a sadistic drill sergeant, the two have each other's backs through long battles in streets blasted by artillery and split by merciless sniper fire. Toward the end, Cowboy is shot by a sniper, and Joker braves a hail of bullets to embrace him before he dies.

This scene is quite moving, and it reflects the finest ideals of unit cohesion in the US Marine Corps. But it is also surprisingly sentimental compared to the ending of Hasford's novella, which takes place in the rugged hills surrounding the US Marine base at Khe Sanh. In Hasford's telling, Joker, Cowboy, and the other members of a small patrol are ambushed by an invisible NVA force. The point man is wounded, and the others must choose between trying to retrieve him, which will result in their being massacred, or abandoning him to save themselves. The situation goes from bad to worse when Joker takes the initiative by blowing his friend Cowboy's brains out.

It is an ugly act that resonates all too well with the title of Hasford's novella: in Vietnam, a "short-timer" was a draftee looking forward to the designated end of his tour. So, the shorter the time, the more likely the soldier will sever the bonds between him and his comrades in arms. Stanley Kubrick was hardly a squeamish director. But this horrifying moment was too strong for his stomach, it seems.

The best film about Vietnam is John Irvin's *Hamburger Hill* (1987), about two divisions of the US Army's Third Battalion trying to seize a mountain ridge in the Central Highlands that is being held by the NVA.

Designated Hill 937 by their officers, the ridge has been dubbed Hamburger Hill by the soldiers, in commemoration of its Korean predecessor, Pork Chop Hill. The final assault, in May 1969, is generally considered to have been a tactical disaster. Equal to its immediate predecessors in terms of vivid action and compelling performances, *Hamburger Hill* is far superior in terms of its portrayal of the war.

To begin, the film highlights the morale-destroying impact of the individual rotation policy, but without trying to equate the soldiers' objections to that policy with political opposition to the war itself. Similarly, we see the headache caused by the arrival of a new batch of FNGs ("fucking new guys"), but we also see how the seasoned unit leader, Adam Frantz (Dylan McDermott), manages to train them on the spot. Further, the film shows how intensely the soldiers hate the order to keep assaulting the hill no matter what. But it does not confound that hatred with any political view. Instead, the soldiers' bitterness toward the ineptness of the "brass" is offset by their resentment toward anti-war protestors who hurl insults at them. And in one scene, Frantz tells an arrogant TV reporter that he has more respect for the NVA than for the media.

I could go on. *Hamburger Hill* does not divide its cast of characters into good guys and bad, depending on their politics. Nor does it automatically ascribe anti-war sentiments to black soldiers objecting to racist treatment in the military. But by now, my drift should be clear: with this superb, powerful film, Hollywood finally got Vietnam right.

Lessons Learned

"Lessons learned" is military talk for not making the same mistake twice. There are two lessons to be drawn from this history. The first is that unit cohesion is not an adequate substitute for a genuine casus belli. When the bullets are whizzing past, it may be dramatically necessary (and sociologically accurate) to show soldiers putting comradeship before cause. But the two are not that easily separated. At some point the shooting stops, and the soldiers ponder why they fight. If no adequate reason presents itself, they grow less willing to walk back into hell.

The second lesson, not yet learned, is that the Washington-Hollywood Pact needs repair. As I have been arguing, the depiction of war in Hollywood films changed dramatically when America's hot war against Fascism ended and its Cold War against Communism began. The original terms of the pact called for the government to support the production and export of Hollywood films in exchange for an assurance that "the pictures distributed abroad will reflect credit on the good name and reputation of this country and its institutions." By the end of the Vietnam era, this two-way covenant in support of both patriotism and profit had devolved into a one-way extortion devoted exclusively to profit.

Today, this is even more the case. Indeed, when it comes to the export of American films to the rest of the world, Hollywood and Washington both act as though it were still 1915 and the film industry was "a business, pure and simple." Since the Cold War ended, there have been numerous attempts by France, Canada, and various allies to carve out a "cultural exception" to the free-trade ethos of the World Trade Organization. To these efforts Washington's response has never wavered: all such claims are illegitimate, because there is no meaningful distinction between trade in commodities and trade in cultural expression. To create a film is the same as to manufacture a widget.

In America, it took a lengthy legal battle to redefine film as a form of artistic expression deserving of First Amendment protection. And the result is a greater degree of creative freedom for post-1960s Hollywood than for any other film industry on Earth. All the more ironic, then, to see both partners to the pact agreeing supinely to let the Chinese Communist Party assume ever more control over the production, content, and distribution of American films in China. In the cogent summary of a recent report from PEN America: "Hollywood is one of the world's most significant storytelling centers, a cinematic powerhouse whose movies are watched by millions across the globe. And yet the choices it makes, about which stories to tell and how to tell them, are increasingly influenced by an autocratic government with the world's most comprehensive system of state-imposed censorship."[21] Today, of course, audiences everywhere are migrating away from theaters and broadcast outlets toward online streaming. But the digital age is not going to restore freedom of expression to Hollywood. Chinese film companies no longer need US help in producing domestic entertainment, including jingoistic war films in which American mercenaries (not soldiers—yet) are cast as the enemy. And Chinese tech companies have all the resources they need to combine social media and streaming services with state-of-the-art censorship, surveillance, and propaganda.

Faced with this situation, Washington and Hollywood have a choice. They can continue to sacrifice their independence for an ever-shrinking share of the Chinese market. Or they can do humanity a favor and reaffirm the artistic freedom that has long been the source of their global appeal. From my vantage point as an admirer of Hollywood's imperfect but creative efforts to portray the hottest and deadliest conflicts of the Cold War, I recommend the latter.

Acknowledgments

Thanks to Ying Zhu, Kenneth Paul Tan, and editors of *Global Storytelling* for the opportunity to contribute to this important issue.

21. James Tager with Jonathan Landreth, *Made in Hollywood, Censored by Beijing* (New York: PEN America, 2021), https://pen.org/report/made-in-hollywood-censored-by-beijing/.

Bomb Archive

The Marshall Islands as Cold War Film Set

ILONA JURKONYTĖ

Abstract

This essay offers a decolonial analysis of the inaugural moment of the United States' Cold War project—the nuclear weapon "testing" in oceanic environments. As an alternative to the usual framing of Pikinni Atoll as a site of the Cold War arms race that tends to invisibilize Marshallese experiences through a Cold War binary logic, this article invites the reader to focus on the Pikinni Atoll as a film set. It offers such an approach with the hope of reframing questions of justice and recognizing the worlds lost due to the production of US nuclear modernity.

Keywords: bomb archive, audiovisual deterritorialization, audiovisual Cold War epistemologies, nuclear colony, nuclear weapons' "testing", extraction through image

By analyzing the founding moment of the US nuclear "testing" in the Marshall Islands, as it is narrated in newsreels produced in 1946, I focus on how audiovisual technology takes part in the production of injustices in the Marshall Islands. The perpetuation of these impacts derives from the US Cold War production. With the US nuclear weapon project, a distinct audiovisual archive gets instigated. Its creation and preservation result in a unique type of injustice that I attribute to a nuclear colonial condition due to inextricable functionality of recordings of this extraterritorial nuclear project. I apply a media-analytic lens to expose a particular type of colonial violence and to offer possible venues for justice claims.

https://doi.org/10.3998/gs.2536

57

In the context of nuclear weapon production, I ask, what does taking a film camera out to sea mean in relation to the notions of territorial (un)making and production of evidence? In other words, what are the implications of the extraterritorial bomb archive? Military presence in the ocean space relentlessly shapes the notion of territory: it percolates practices of extraction and contributes to shaping knowledge-production, especially when it comes to science and area studies in humanities. I emphasize the importance of recognizing the role of media in historical and contemporary sea-bound conflicts such as legislation of national and international waters, territory-making, border crossing, and military weaponry "testing" at sea. My research is driven by a hope that interrogating the past and present from an oceanic perspective, in combination with an analytic film industry lens, can help bring a necessary shift from a Cold War binary framing of the global order (as defined by tensions between the two superpowers) to a nonpolarized demand for accountability.

While I take into account the context of struggle for self-determination and acknowledge the ongoing spatiotemporal complexities defining a non-homogeneous Marshallese community, my research is not meant to directly comment on these complexities. For this purpose, I draw on the latest and most pertinent ethnographic research of Sasha Davis, Barbara Rose Johnston, Jessica A. Schwartz, and others.[1] This important work informs my

---

1. Sasha Davis, *The Empires' Edge: Militarization, Resistance, and Transcending Hegemony in the Pacific* (Athens: University of Georgia Press, 2015); Interviews with Bikinian Elders, Bikiniatoll.com, accessed May 18, 2001, http://www.bikiniatoll.com/interviews.html, in Ruth Levy Guyer, "Radioactivity and Rights," *American Journal of Public Health* 91, no. 9 (September 2001): 1375; Jeffrey Sasha Davis, "Scales of Eden: Conservation and Pristine Devastation on Bikini Atoll," *Environment and Planning D: Society and Space* 25, no. 2 (April 2007): 213–35; Steve Brown, "Poetics and Politics: Bikini Atoll and World Heritage Listing," in *Transcending the Culture-Nature Divide in Cultural Heritage*, vol. 36: *Views from the Asia–Pacific Region*, ed. Sally Brockwell, Sue O'Connor, and Denis Byrne (Canberra: Australian National University Press, 2013), 35–52; Barbara Rose Johnston, "Nuclear Disaster: The Marshall Islands Experience and Lessons for a Post-Fukushima World," in *Global Ecologies and the Environmental Humanities Postcolonial Approaches*, ed. Elizabeth DeLoughrey, Jill Didur, and Anthony Carrigan (New York and London: Routledge, 2015), 140–61; Barbara Rose Johnston and Holly M. Barker, *Consequential Damages of Nuclear War: The Rongelap Report* (London: Routledge/Taylor & Francis, 2017);

research and allows me to develop my argument on how the production and circulation of the bomb archive is at the core of nuclear colonial injustices.

Complex intertwining of military-scientific as well as visual production took place in unprecedented oceanic nuclear "tests," which in the Pacific started with Operation Crossroads in 1946. The scale and range of the US military-scientific productions in the Pacific were well beyond ordinary. From the very start, while unprecedented international agreements permitted the United States to detonate never-before fission yields, concurrently another important world record was being set. More than half the world's supplies of film stock and around eighteen tons of cinematography equipment were brought to Pikinni Atoll in the Marshall Islands.[2] Operation Crossroads rendered what until then had been audiovisually a nearly undocumented place into one of the world's most photographed at the time. The US cinema-military complex deployed over five hundred cameras to document Operation Crossroads.[3] An immense archive of moving images has been harvested from these highly radioactive, destructive explosions.

The goal of this essay is to contribute to decolonial perspectives by exposing how the US cinema-military complex, after taking shape in World War II,[4] continued its unscrupulous role not only in inventing the Cold War through its oceanic cinematic operations but also in becoming an important tool used to expand oceanic spheres of direct US influence and produce pervasive material

---

Jessica A. Schwartz, "Marshallese Cultural Diplomacy in Arkansas," *American Quarterly* 67, no. 3 (September 2015): 781–812; Jessica A. Schwartz, *Radiation Sounds: Marshallese Music and Nuclear Silences* (Durham, NC: Duke University Press Books, 2021); Robert Stone, dir., *Radio Bikini* (New York: Robert Stone Productions, 1988).

2. I use the Marshallese (also known as Ebon) transliteration Pikinni in order to distinguish the name of the place from militarized and gendered connotations of the German, French, and American transliteration of Bikini.

3. Kevin Hamilton and Ned O'Gorman, *Lookout America! The Secret Hollywood Studio at the Heart of the Cold War* (Hanover, NH: Dartmouth College Press, 2019), 74.

4. For more on the functionality of the military's cinema complex inside the US military and US military's contributions toward development of portable film-exhibition technology, see Haidee Wasson, "Experimental Viewing Protocols: Film Projection and American Military," in *Cinema's Military Industrial Complex*, ed. Haidee Wasson and Lee Grieveson (Berkeley: University of California Press, 2018), 24–43.

and symbolic legacies. I argue that the cinematic apparatus played an important part in the production of the US Cold War and continues its extensive role in the dynamics of Cold War legacies. The production of the Cold War to this day takes a toll on communities in the Pacific Islands in several ways now recognized and in others not yet widely acknowledged. In pursuit of shifting the usual framing of Pikinni Atoll through Cold War categories that are productive of US nuclear modernity and its fetishizing tendencies, I propose a decolonial analysis of how image production, circulation, and its archiving practices are at the heart of ongoing violence. My focus is on this archive's production and its long-term cultural, political, and epistemological effects.

Nuclear Audiovisual Operation

I delineate two major categories of audiovisual documents that were produced in relation to the Marshall Islands nuclear "tests" with a plan to distinguish parts of it for wide distribution: (1) newsreel films from the Marshall Islands and (2) raw footage as documentation of the explosions. Both broad categories contain several but not definitive subcategories:

1. Newsreel films from the Marshall Islands:
    1.1. US military preparations for weapon "tests"
    1.2. Postexplosion assessment
    1.3. Scientists revisiting the Marshall Islands to measure radiation impacts on human and nonhuman bodies
2. Raw footage/documentation of explosions, which has been absorbed into other audiovisual productions, such as:
    2.1. Fiction films
    2.2. Documentary films
    2.3. News coverage
    2.4. Artistic films
    2.5. Music videos

These are general categories summing up a plethora of audiovisual works dating from 1946 onward. I list these broad categories following my goal to analyze the modalities of an epistemic regime that is directly manufactured from (audio)visual[5] documentations of the extraterritorial US nuclear project. I name the variety of images of the US nuclear "test"[6] explosions a *bomb archive,* which I place in relation to other succinct notions that pertain mainly to scientific language, such as *bomb carbon* and *bomb effect.*[7]

An important distinction between the two main categories of the bomb archive is their function. While newsreels were deployed in internal, national, and international communication, raw footage/documentation of explosions were primarily meant for military-scientific assessment of the effects of the bomb.[8] Both these categories and their multifunctional application were invented during Operation Crossroads, and both have a differently violent reach. Earlier documentations of nuclear explosions (July 16, 1945,

---

5. I mark "audio" in brackets in "(audio)visual" every time I refer to the US military's produced moving images of Operation Crossroads because of the alleged equipment failure causing the iconic images of aquatic mushroom clouds recording without an audio track. Ostensibly, the intention to record audio was there, but the blast power interfered with the quality of the material. In the majority of moving-image works representing Operation Crossroads nuclear explosions, the audio part is added in postproduction. In other cases, the image is projected with no soundtrack.

6. In an effort to avoid the normalizing of Cold War language, I propose the use of quotation marks around the word "test" when referring to nuclear weapons explosions. I want to avoid the misleading connotation that these took place in a scientific laboratory environment instead of in outdoor environments that suffer ongoing effects. I do this in line with Kathryn Yusoff's geotemporal critique of conceptualizing the Anthropocene as something that has consequences for human population while extractive practices have long created adverse conditions for racialized individuals and communities. See Kathryn Yusoff, *A Billion Black Anthropocenes or None* (Minneapolis: University of Minnesota Press, 2018).

7. For more on bomb carbon and bomb effect, see Rose Eveleth, "Nuclear Bombs Made It Possible to Carbon Date Human Tissue," *Smithsonian Magazine,* accessed May 25, 2022, https://www.smithsonianmag.com/smart-news/nuclear-bombs-made-it-possible-to-carbon-date-human-tissue-20074710/.

8. For more on the history of the application of cinematic apparatuses in the Marshall Islands, see William A. Shurcliff and US Joint Task Force One, *Bombs at Bikini: The Official Report of Operation Crossroads* (New York: W. H. Wise, 1947); Jack De Ment, "Instruments of Operation Crossroads," *Military Engineer* 39, no. 264 (1947): 414–19.

Trinity; August 6, 1945, Hiroshima; August 7, 1945. Nagasaki) were comparable neither by scale (visual documentation of Operation Crossroads was the biggest of all) nor function (visual documentation of Operation Crossroads was used for a wide array of purposes: from image as science data to image used for internal and international media campaigns).

Through the highly controlled efforts of the US military,[9] the images of Operation Crossroads became part of the shared imaginary and the iconic representation of the nuclear explosions. Kevin Hamilton and Ned O'Gorman explain that this operation was intended to both test the fortitude of naval vessels in an atomic blast and "provide ample images to the U.S. and global publics of America's newly invented weapons (something . . . that both Hiroshima and Nagasaki failed to do). As such, photography was as critical to Crossroads as ballistics and in certain respects more critical, as the U.S. had never set up a photographic operation quite like this, whereas the bomb designs had already been 'proven' in Japan."[10] Robert Hariman and John Louis Lucaites claim that the initial visual representation of atomic warfare applied in Japan was "artistically modest, morally ambivalent, tilted toward abstraction and ethical justification, and not yet anchored in one medium or image."[11] In that sense, Operation Crossroads was altogether different—fully fleshed out and anchored in the cinema-military complex.

Other than being central to highly controlled US public campaigns that were foundational for the international Cold War regime, images of the Pikinni Atoll nuclear weapon explosions enter the public sphere under a guise of neutrality as representations of scientific "tests." As such, they are

---

9. For more information on US communications about nuclear projects, see Beverly Deepe Keever, *News Zero: The New York Times and the Bomb* (Monroe, ME: Common Courage Press, 2004).
10. Hamilton and O'Gorman, *Lookout America!*, 75.
11. Robert Donald Hariman and John Louis Lucaites, "The Iconic Image of the Mushroom Cloud and the Cold War Nuclear Optic," in *Picturing Atrocity*, ed. Geoffrey Batchen et al. (London: Reaktion Books, 2013), 135–46.

separated from their sociomaterial Marshall Islands' context and perform as raw data. Representations of nuclear weapon explosions fluctuate between a scientific register as raw data and a film industry approach as raw material. They both are means for different and yet interconnected ends: scientific meaning and aesthetic meaning.

The aspects of scale of the nuclear visual operation were "tested," assessed, and advanced during Operation Crossroads. This first oceanic extraterritorial nuclear "test" forecasted the need for a separate branch of institutional coordination inside the internal organization of the US military. As Hamilton and O'Gorman put it, "The U.S. government had to manage not only a technological system and its biochemical artifacts, but also the collection of images, stories, and data that inherently threatened to upset America's place within a precarious post-war international order."[12] They observe that, back in the 1940s, the excessively broadcasted images of Operation Crossroads lacked a compelling narrative—it had only a palette of images showing the awful might of atomic explosions.[13] "The story of Crossroads threatened to become the story of American military recklessness. The Navy tried to avert this public relations fiasco by making the images, and indeed the cameras themselves, the story."[14] Today, we can recognize that success was attained in the effort to perpetually fetishize technological image-production aspects, which, throughout the second half of the twentieth century and beyond, are most often framed as "American nuclear modernity." Nuclear technological achievement and demonstrations of power to the USSR were crafted as the main narratives of the US Cold War. Furthermore, information about the oceanic nuclear weapon "tests" by US Cold War strategists and communicators has reached the level of cliché—the representation of an atomic mushroom cloud started signifying the Cold War itself. Iconography of the oceanic mushroom cloud together with the bikini bathing suit became

---

12. Hamilton and O'Gorman, *Lookout America!*, 74.
13. Hamilton and O'Gorman, 78.
14. Hamilton and O'Gorman, 76.

major tools through which the Marshall Islands was represented. Teresia K. Teaiwa's feminist decolonial critique of the bikini bathing suit as a gendered obscuring device[15] of Indigenous subjectivity is my inspiration for looking for methodological frameworks that would permit recognition of the extensiveness of obscuring agendas. I recognize such a perseverant, obscuring framework in the audiovisual bomb archive.

Operation Crossroads served as an initial merger between the military-scientific and cinema-military complex. It produced a logistical prototype to the Lookout Mountain Laboratory, which was established right after Operation Crossroads and provided growing infrastructure around the bomb archive's production, preservation, management of access, and circulation. According to Hamilton and O'Gorman, Lookout Mountain Laboratory, also known as Lookout Mountain Air Force Station, served as the headquarters of the 1652nd Motion Picture Squadron of the US Air Force from 1947 to 1969.[16] After implementing some variably successful documentation of Operation Crossroads, based on an acknowledgment of the importance of visual accounts and an understanding of the logistical challenges of the scale of such a cinematographic operation, Lookout Mountain Laboratory became, arguably, "the Cold War's most prolific and influential film studio."[17] While balancing between imagery production and imagery archiving, between utter secrecy and receiving a nomination for an Oscar from the Academy for Motion Arts and Sciences, Lookout Mountain Laboratory was also registering patents and publishing papers in technical journals dedicated to sound recording and scientific photography.[18] As much as being remarkably prolific in image production, Lookout Mountain was

---

15. Teresia K. Teaiwa, "Bikinis and Other S/Pacific N/Oceans," in *Militarized Currents: Toward a Decolonized Future in Asia and the Pacific*, ed. Setsu Shigematsu and Keith L. Camacho (Minneapolis: University of Minnesota Press, 2010), 15–32.

16. Lookout Mountain Laboratory, "Lookout America!," Archives of the 1352nd Motion Picture Squadron, accessed January 17, 2022, https://www.lookoutamerica.org/.

17. Lookout Mountain Laboratory.

18. Lookout Mountain Laboratory.

not the only studio engaged in bomb archive production.[19] When thinking about this moment, it is important to acknowledge not only the scale but also the role of the extractive operation in the Marshall Islands in areas of knowledge-production: inventions in audio and visual recordings, science photography, bomb-carbon facilitated forensics, and outer space research, as well as the fact that among a plethora of unprecedented scientific productions, an entire discipline of ecosystem ecology derives from the Pacific nuclear weapon "tests."[20] These types of knowledge together with an abundant iconization of the mushroom cloud[21] were produced through an overarching system of data classification that functions not only by closing access to data but by applying differential logic to providing access to selected data. There is no doubt that the nuclear weapons industry together with its derivative research play a substantial role in the world's economy.

Ocean and Occupation without Possessing

As evidenced by Lookout Mountain Laboratory's operations, the paradox of invisibilizing through hypervisibility is produced in the conjuncture of the military-scientific and cinema-military industrial complexes. To bypass the US Cold War–produced fetishizing that is usually applied in analysis of US nuclear representations, I propose a critical oceanic and film-industry lens to embrace questions of justice in the critique of hypermilitarization and recognition of worlds lost due to the production of US nuclear modernity.

---

19. For more information on the early stages of production of nuclear weapons representations, especially in settler-colonial contexts, see Susan Courtney, "Framing the Bomb in the West: The View from Lookout Mountain," in *Cinema's Military Industrial Complex*, ed. Haidee Wasson and Lee Grieveson (Berkeley: University of California Press, 2018), 210–26.
20. Joel Bartholemew Hagen, *An Entangled Bank: The Origins of Ecosystem Ecology* (New Brunswick, NJ: Rutgers University Press, 1992); Laura J. Martin, *Wild by Design: The Rise of Ecological Restoration* (Cambridge, MA: Harvard University Press, 2022).
21. For more on the history of iconization of the mushroom cloud, see John O'Brian and Art Gallery of Ontario, *Camera Atomica* (Toronto: Art Gallery of Ontario, 2015).

Historical examples of neocolonial Cold War employments of the ocean space are plentiful. As early as August 1946, the United States began efforts toward including floating ice masses in its national security discourse, which led to attempts to occupy frozen water masses as land.[22] The physical feature of water taking solid shape in low temperatures led to juridical confusion. However, the fact that the US military focused on it meant that efforts were put in place to reconcile juridical contradictions and use this to allow an exemption of ocean space from selective aspects of national and international law.[23] Bruun and Steinberg's research shows how, between 1952 and 1978, the floating ice mass became part of what they call the "wider U.S. techno-political network of knowledge production that spanned across the Arctic and beyond."[24] According to Bruun and Steinberg, both scientists and the military were struggling with the same set of questions: "How could an environment of shifting mobile solid water that could be 'occupied' but not 'possessed' be assimilated into a system of spatial organization that assumes divisions between solid and liquid, between land and water, and between 'inside' (territory) and 'outside' (non-territory)?"[25] The physical features of water freezing into large shelves, contamination patterns, oceanic ecosystems growing reefs, and islands forming and disappearing challenge terrestrial-based juridical systems. These challenges demonstrate how the familiar (and usually undisputed) notions of what constitutes political, economic, juridical, and symbolic power over territories do not seem adequate when this power shifts from land mass to water. And yet, as both history and present-day political events demonstrate, many of the most intense and consequential of these struggles take place precisely over—not to mention on and under—water. The stakes in these queries

---

22. Johanne Bruun and Philip E. Steinberg, "Placing Territory on Ice: Militarisation, Measurement and Murder in the High Arctic," in *Territory Beyond Terra*, ed. Kimberley Peters, Philip Steinberg, and Elaine Stratford (London and New York: Rowman & Littlefield, 2018), 147–65.
23. Bruun and Steinberg, "Placing Territory on Ice."
24. Bruun and Steinberg, 147.
25. Bruun and Steinberg, 149.

tend to appear abstract when territorialization by superpowers does not expose human inhabitants, thus appearing free of political implication.

Such occupation without possession in the ocean space seems to be prevalent in the postwar US context. Technology historian Ruth Oldenziel calls this phenomenon the "deterritorialization of power"[26]—expansionism without juridically registrable and materially visible evidence. Oldenziel identifies the technoscientific modus operandi as leaching onto juridically "thin" places by introducing technological volume, or technological "thickness," as she puts it—to these places. In Oldenziel's analysis, such juridically "thin" and technologically "thick" places are islands scattered around the globe. I find this impulse for instituting extractivist practices on ice, corals, and atolls comparable and eye-opening. While the 1946 discovery of the T-1 iceberg was classified as a military secret,[27] the first oceanic nuclear weapon "test" (Operation Crossroads) was in the same year constructed as a public event. Both instances were driven by a similar impulse for occupying without possessing. The image-production complex was deployed in the case of Operation Crossroads in unprecedented ways, and this allows a glimpse into technological layering in juridically thin places from a perspective of visual culture analysis.

Oldenziel emphasizes that the United States rules over extensive—but invisible to its citizens—island possessions: the Commonwealth of Puerto Rico, Guam, American Samoa, Johnston Atoll, Navassa Island, Micronesia, Marshall Islands, the Commonwealth of the Northern Marianas, Palau, and the US Virgin Islands of St. Thomas, St. John, and St. Croix. Oldenziel notes that these US territories are the largest colonial holdings in the (post) colonial era,[28] exceeding the combined population of the overseas territories

---

26. Ruth Oldenziel, "Islands: The United States as a Networked Empire," in *Entangled Geographies: Empire and Technopolitics in the Global Cold War*, ed. Gabrielle Hecht (Cambridge, MA: MIT Press, 2011), 13–41.

27. Bruun and Steinberg, "Placing Territory on Ice," 147.

28. By bracketing (post) in (post)colonial, I make reference to Ann Laura Stoler's emphasis on durabilities of colonial presence. See Ann Laura Stoler, *Duress: Imperial Durabilities in Our Times* (Durham, NC: Duke University Press, 2016), x.

of Britain and France.[29] I argue that what sets the Pikinni Atoll apart from all of these other instances of United States' holdings is that it is (1) the place of invention of what is termed a *nuclear colony*, (2) it is part of an important moment in the production of US island-networked power, and (3) it is the birthplace of the most public bomb archive—an archive central to Cold War securitization discourse and practice. And as such it continues to be interpreted through Cold War epistemic categories that are incapable of grasping the current conditions of deterritorialized colonial extraction that the United States exercises across the islands while the islanders are perpetually striving for decolonization.

It is important to note that the first oceanic nuclear weapons "test" was the first to be conducted by the United States outside of its own territory. The second, and best visually documented, oceanic explosion was immediately dubbed the world's first nuclear disaster.[30] The questions of disaster and environmental crisis[31] are inseparable from all stages of nuclear energy production—such as the nuclear disasters at Chernobyl, Fukushima, and more—given that nuclear pollution remains toxic for tens of thousands of years.[32]

The time of inception of the Cold War is replete with events that at that time were unprecedented. Paradoxically, the solidifying of the Cold War was achieved through material and juridical maneuvers in the vast and fluid space of the ocean. The UN Trusteeship Agreement[33] with the United States

29. Oldenziel, "Islands," 13–41.
30. Jonathan M. Weisgall, *Operation Crossroads: The Atomic Tests at Bikini Atoll* (Annapolis, MD: Naval Institute Press, 1994), ix.
31. Johnston, "Nuclear Disaster."
32. For more on varieties and the longevity of nuclear pollutants, see Eric Semler, James Benjamin, and Adam Gross, *The Language of Nuclear War: An Intelligent Citizen's Dictionary*, 1st ed. (New York: Perennial Library, 1987); Keever, *News Zero*; Johnston, "Nuclear Disaster"; Barbara Rose Johnston and Holly M. Barker, *Consequential Damages of Nuclear War*; Schwartz, *Radiation Sounds*.
33. Susan Kurtas, "U.N. Documentation: Trusteeship Council: Strategic Trust Territory of the Pacific Islands," research guide, United Nations Dag Hammarskjöld Library, New York, accessed August 12, 2021.

in 1947 put the Marshall Islands, the Caroline Islands, and the Mariana Islands under a unique type of US control. This extraordinary juridical formation legitimized military-scientific and media operations.

Other than the fact that the Trusteeship Agreement vastly expanded the oceanic presence of the United States, it also allowed it the right to close any areas of this "strategic" territory at any time "for security reasons." In 1954, legal scholar Emanuel Margolis summed it up as follows: "Upon United States insistence, the entire territory—comprising ninety-eight distinct islands and island units with a combined land area of 846 square miles, spread over some three million square miles of ocean—was set up as 'strategic' under Article 82 of the U.N. Charter."[34] The Trusteeship Agreement allowed the United States to expand its control over lands and waters three times the size of US territory. At that time, well-known[35] detrimental effects on human and nonhuman lives led Margolis to conclude as early as 1954[36] that "the laws of humanity suggest and the law of nations requires immediate cessation of the thermonuclear experiments in the Pacific Proving Grounds."[37] The solution he offered, however, should the "testing" not stop entirely, was to move it to the remote Arctic region. Sadly, this is the same argument that led the United States to the Marshall Islands in the first place. Today, with the knowledge we have about the glaciers melting, we can understand how the perception of oceanic remoteness (as separate, nonconnective, isolated, and empty) was at the heart of false convictions that propelled twentieth-century nuclear weapons production. Despite warnings from experts and scholars, the United States proceeded with active "testing" in the Marshall Islands till 1958.

---

34. Emanuel Margolis, "The Hydrogen Bomb Experiments and International Law," *Yale Law Journal* 64, no. 5 (1955–1954): 630.
35. The harmful effects of radiation exposure were already known as early as the 1920s and 1930s. For more on this topic, see Gabrielle Hecht, *Being Nuclear: Africans and the Global Uranium Trade* (Cambridge, MA: MIT Press, 2012).
36. Margolis's reaction could have been motivated by another well-documented audiovisually nuclear process, the Castle Bravo thermonuclear "test," which took place March 1, 1954.
37. Margolis, "The Hydrogen Bomb Experiments and International Law," 647.

Extraction through Image

Extraction through image takes place in scientific laboratories and in public communication. Such extraction begins at the point of image production and extends through its circulation. The image and its analysis are crucial in the process of refining nuclear weapons. Representation of nuclear technology is both scientific (as a means of analysis in order to measure the effects of explosions[38] and the effects of radiation on human[39] and nonhuman[40] bodies) and symbolic (as a means of shifting world power balance). I call this nonconsensual scientific and symbolic production *extraction through image.*

To understand the ways in which violence operates through the sphere of the (audio)visual, it is important to analyze the production of the bomb archive. Image production is a defining factor in the constellation of encounters between parties of the Trusteeship Agreement. When tracing how nuclear weapons explosions in the Marshall Islands started, the timeline itself appears very rushed for an operation of such magnitude. The scale and speed with which half of the world's film stock, eighteen tons of cameras, and at least 412 cameramen[41] were sent to this remote and seemingly difficult-to-reach island speak for themselves. Looking back at the sequence of events, it is important to notice the large degree to which the US military focused on (audio)visual documentation of this moment. In such context, the Atolls, the Marshall Islands, and the entire territory of the Trusteeship Agreement[42] (among other formations) were turned into a film production set. The logic of film production dominated throughout the duration of this military operation. Aspects of footage production, its meaning, and

---

38. Shurcliff and US Joint Task Force One, *Bombs at Bikini.*
39. Yusoff, *A Billion Black Anthropocenes or None*, 46.
40. Susan Schuppli, "Radical Contact Prints," in *Camera Atomica*, ed. John O'Brian (Toronto: Art Gallery of Ontario, 2015), 278–91.
41. Hamilton and O'Gorman, *Lookout America!*
42. The Marshall, the Mariana, and the Caroline Islands.

circulation demonstrate how film production in the Marshall Islands is part of nuclear violence.

Historical records suggest that from the moment the United States took over the territory of the Marshall Islands it claimed ownership. Even if the Trusteeship Agreement did not grant ownership itself, it granted governing freedoms that are often associated with ownership. Representatives of the United States did not shy away from testing the limits of this exceptional trust(eeship). The US government insisted that the Trust Territory of the Pacific Islands (TTPI) be designated a "strategic trust"[43]—one in which the administering power had a national security interest. The status of strategic trust meant that the United States would be accountable only to the UN Security Council, where the United States held veto power, rather than to the UN General Assembly, which administered all other trust territories.

The chronology of US actions after its military entered the waters of the Marshall Islands in 1944 demonstrates how quickly infrastructure for explosions and for setting up the media operation developed. It is apparent that the nuclear weapons "test" site intersected with that of a film-set. Explosions had to be visible: frameable, well-lit, with sufficient openings for camera angles. Cameras and microphones had to be sheltered from winds and blast power. The concrete structures were built specifically for (audio)visual documentation. A choreography of aerial shots was planned in coordination with cameras. This is especially true of the second Baker "test," representations of which were massively reproduced.

One of the major signs of asymmetrical colonial relations is the right to land. US officials removed inhabitants from the lands and waters. While Steinberg described Micronesian Islanders' regard for water as a "land-like space of distinct places,"[44] and the survivors of nuclear violence assert

---

43. Schwartz, "Marshallese Cultural Diplomacy in Arkansas."
44. Philip Steinberg, *The Social Construction of the Ocean* (Cambridge and New York: Cambridge University Press, 2001), 43.

their exceptional relation to the place,[45] the US military treated the ocean as colonially "asocial"[46]—an empty space. Sasha Davis describes how the "emptiness" of the Marshall Islands was initially identified but also further emphasized as a major resource that legitimized nuclear operations.[47] Deceiving the islanders about the scale and longevity of the operation is comparable to theft of waters and the islands.[48]

The islanders' desire to stick to the initial agreement and return to their islands in the 1970s posed another opportunity for extraction—knowledge about the effects of radiation was produced without initial consent and by withholding information about it from the islanders themselves. Imaging technology played a crucial role in this process. After several attempts by islanders to return to their ancestral lands, it was generally agreed that the nuclear explosions eliminated the possibility of permanent, sustainable return. Davis calls such colonial settlements, which are being significantly changed by the US military, *baseworld*.[49] The Marshall Islands epitomize this, along with several other colonial modalities. These modalities are unique to the Marshall Islands in their intense combination but at the same time they are exemplary of broader tendencies of the United States' "deterritorialization of power" through island-driven, neocolonial expansion. Still today, according to the United Nations' plan for ongoing decolonization and 2020 statistics, the majority of nonself-governed communities inhabiting islands around the globe amounts to around two million individuals.[50]

---

45. Interviews with Bikinian Elders, Bikiniatoll.com, accessed May 18, 2001, http://www. bikiniatoll.com/interviews.html, in Ruth Levy Guyer, "Radioactivity and Rights," *American Journal of Public Health* 91, no. 9 (September 2001): 1375.
46. Steinberg, *The Social Construction of the Ocean*.
47. Davis, *The Empires' Edge*.
48. For theft and property relations in settler-colonial contexts, see Robert Nichols, *Theft Is Property! Dispossession and Critical Theory* (Durham, NC: Duke University Press, 2020).
49. Davis, *The Empires' Edge*.
50. "The United Nations and Decolonization: Past to Present," United Nations, accessed August 1, 2021, https://www.un.org/dppa/decolonization/en.

Audiovisual Deterritorialization of Power

Elimination of Indigenous societies is at the center of Patrick Wolfe's definition of a settler colony.[51] Although Pikinni Atoll shares this characteristic of a settler colony, when compared with studies of other settler colonies, the core constitutive elements of the Marshall Islands follows a different causal sequence. In the Marshall Islands, elimination of the Indigenous population from the lands was not a means of extraction but an immediate condition. Removal of the population was facilitated by post–World War II international legislation regarding the island states. Unlike other settler-colonial contexts, the supposed return of stolen lands, or lack of relocation of populations from the danger zone, such as in the case of Roñḷap Atoll, was used as another occasion for extraction: the US military-scientific complex studied the post-"test" environment, including the islanders' physical bodies, to observe the long-term impact of nuclear radiation.[52] As DeLoughrey claims, "Despite the excessive surveillance and documentation of their radiogenic illnesses, to this day the majority of affected islanders have been refused access to their medical records and have inadequate medical treatment."[53] The quest for the withheld records is ongoing[54] and once in a while takes the shape of demands by the public that all classified documents pertaining to "testing" be released to the Marshallese.[55] The effects not only of the radiation itself but also of the withheld knowledge[56] are felt to this day and

51. Patrick Wolfe, "Settler Colonialism and the Elimination of the Native," *Journal of Genocide Research* 8, no. 4 (December 2006): 387–409.
52. Yusoff, *A Billion Black Anthropocenes or None*; Johnston, "Nuclear Disaster"; Jonathan M. Weisgall, "Statement of Jonathan M. Weisgall Legal Counsel to the People of Bikini Before the House Natural Resources Committee," US Department of Energy, Office of Scientific and Technical Information, February 24, 1994.
53. Elizabeth M. DeLoughrey, "The Myth of Isolates: Ecosystem Ecologies in the Nuclear Pacific," *Cultural Geographies* 20, no. 2 (2013): 178.
54. Seiji Yamada and Matthew Akiyama, "'For the Good of Mankind': The Legacy of Nuclear Testing in Micronesia," *Social Medicine* 8, no. 2 (January 2013): 83–92.
55. Schwartz, "Marshallese Cultural Diplomacy in Arkansas," 793.
56. For more on the impact on health and cultural practices of the Marshallese communities

should be understood as nonconsensual extraction, which is part and parcel of the same data classification system as the bomb archive.

To most observers, the oceanic nuclear colony is primarily a mediated experience. Audiovisual reproducibility of the invention of the nuclear colony points to a processual definition of it as a structure of a settler colony (drawing on Wolfe's emphasis on a settler colony as an ongoing structure rather than a temporally circumscribed event). I call this mediated structure of a settler colony an *audiovisual deterritorialization of power*, and a source of injustice.

Like many island nations, the Marshall Islands' case shows characteristics of both early-day colonialism and twentieth-century modes of extraction. A major distinctive feature of the Marshall Islands is the fact that image production is at the core of the invention of its particular nuclear colonial modality. The Marshall Islands case is exceptional in that it witnessed the capacity of camera and broadcast attention[57] and served as a building block for technopolitical power, as well as a key to understanding how these new modes of power operate.

Following independence on May 1, 1979, the Marshall Islands became a sovereign republic. And the US civilian population has never actually settled on Marshall Islands territory. Nevertheless, the ultimate elimination of agency from representation in the context of ongoing land dispossession can be seen as a particular type of settler colony. This type of settler colony has not historically been attributed to its territorial ambitions due to the vastness of its surrounding waters, which are not entirely registrable through a geopolitical framework. To a large extent, this is due to spatially and temporally divergent legislation. A nuclear colony was, thus, invented and maintained through a combination of nuclear, oceanic, and imaging extractive practices.

---

caused by nuclear "testing" and subsequent withholding of medical data, see Johnston, "Nuclear Disaster"; Johnston and Barker, *Consequential Damages of Nuclear War*; Schwartz, *Radiation Sounds*.

57. Schuppli, "Radical Contact Prints," 280.

I call the momentary and perpetual visibility of the nuclear bomb archive—the former as a "special effect" on a film set and the latter as a canonized immortality of recording—the *audiovisual deterritorialization of power*. This audiovisual deterritorialization of power signifies pretend visibility of action. Due to its extensive post–World War II visibility, the Marshall Islands are unique compared to other nuclear colonies, but at the same time, its invisibilizing visibility plays a part in overcasting imaginaries on broader nuclear-colonial modalities.

The Trusteeship Agreement between the United Nations and the United States was enacted as a political and a media/cinematic process. Before the Trusteeship Agreement was signed on April 2, 1947, the Pikinni Atoll in 1946 saw the simultaneous production of film sets for (1) an open-air science "laboratory" and (2) narrative film production. Both types of audiovisual productions took place in parallel, and both furthered political goals. Universal Studios, in collaboration with the US military, produced several newsreels prior to the "tests." One of them, "Ready for Atom Tests at Bikini,"[58] depicts the moment of a supposed agreement between US military officials and the Pikinni Atoll population to resettle. The camera registers the apparent consent of the Pikinni community to leave the Atoll. There are no negotiations, no questions asked—only a docile, supposed agreement to leave the Atoll to the US military's care. As the Marshallese people continue reentering their reparation claims in the courts, revisiting the moment of "agreement" is pertinent. As well, medical records are continuously being withheld while the Compact of Free Association between the United States and the Marshall Islands is coming to an end in 2023.[59]

Peter Hales reveals that the newsreel "Ready for Atom Tests at Bikini" was produced not from documentary footage but of scripted[60] material for

---

58. "Ready for Atom Tests at Bikini," in Peter B. Hales, *Outside the Gates of Eden: The Dream of America from Hiroshima to Now* (Chicago: University of Chicago Press, 2014,) 27.
59. Office of Insular Affairs, "Compacts of Free Association," US Department of the Interior, October 15, 2015, https://www.doi.gov/oia/compacts-of-free-association.
60. Hales, *Outside the Gates of Eden*, 27.

which a group of Pikinni Atoll inhabitants were asked to perform a meeting with a single US military representative. Hales follows the chronology of the Pikinni Atoll events from January 10, 1946, when US president Harry Truman officially licensed the bombing. The footage showing the supposed willingness of islanders to abandon their home environment was filmed a month after the agreement supposedly took place. Hales investigated the raw footage dating a week before evacuation and almost a month after the declared date—March 3—instead of the official date February 10, 1946. Hales reveals that the raw footage that did not make it to the final cut exposes fabrication beyond the pretend date. In left-out footage, the Atoll chief Juda gets upset about being forced to repeatedly perform consent for camera. He stands up and walks directly to the cameraman just to say into the camera, "All right; is that all?"[61] Then he is seen storming off the set. I identify this as a moment that Nicholas Mirzoeff calls "countervisuality," an insertion of "the right to look," which he defines as "requiring the recognition of the other in order to have a place from which to claim rights and to determine what is right. It is the claim to a subjectivity that has the autonomy to arrange the relations of the visible and the sayable."[62] This Marshallese claim of the right to look is in the archive vaults, awaiting restoration and transfer to digital formats. The archival imperative and the protection of US military archives through classification practices have prevented the raw footage from being used as evidence in a counternarrative. Meanwhile, most of the online sources offer falsified narratives from military-produced versions.

The film production team delayed[63] the evacuation of the islands for one week in order to get the required quality of performance from their untrained "actors." In the context of a very rushed time line, this further highlights the importance of film production in the framework of the nuclear operation. Pikinni Atoll became a film set with rehearsals, main

---

61. Hales.
62. Nicholas Mirzoeff, *The Right to Look: A Counterhistory of Visuality* (Durham, NC: Duke University Press, 2011), 1.
63. Hales, *Outside the Gates of Eden*, 28.

protagonists, extras, repeated takes, and multiple angles. There was film direction and a clear vision of how this performance would steer the viewers' response. However, I believe this filmic instance should be approached not merely as a cinematic performance but rather as a binding document, especially knowing that the written treaty between Pikinni Atoll inhabitants and the US government does not exist—only the ubiquitous audiovisual edits of the islanders agreeing to leave.

The US military's attention to audiovisual production was rather exceptional, valued not as a documentary but as a document. This might be the first and the last document of its kind produced once film became widely embraced by the military following the speedy World War II deployment of cinematic combat functionality.[64]

The edited footage used in "Ready for Atom Tests at Bikini" is not just a straightforward piece of propaganda but is also a source of one of the most quoted alleged facts about Pikinni Atoll, emphasizing how the islanders left their homeland willingly. This story comes from the newsreel itself, and it had been repeated in many written sources, tending to reappear as authentic documentation in multiple other audiovisual productions. However, when studying the subject more deeply, one is left with the impression that the Marshallese had not much of a choice. And yet, the insistence on this move as an informed and voluntary act haunts almost every source.

Information about the islanders' decision to leave Pikinni Atoll needs closer analysis. The moment of agreement is often read as a separate instance. Johnathan M. Weisgall writes that the islanders' decision to leave their environment was not based solely on a desire to see mankind benefit from nuclear "testing." Since the defeat of Japan, US ships were bringing food, other supplies, and medical officers who provided free services: "By the end of 1945 the Americans had built a store, an elementary school, and a medical dispensary on the atoll."[65] Weisgall acknowledges that saying "no" to US officials did not

---

64. Wasson and Grieveson, *Cinema's Military Industrial Complex.*
65. Jonathan M. Weisgall, "The Nuclear Nomads of Bikini," *Foreign Policy*, no. 39 (1980): 78.

seem like a choice. As people were preparing to leave for an evacuation site (temporarily, they thought, for two weeks), the first of two hundred and fifty vessels, one hundred and fifty aircraft, and forty-two thousand military and scientific personnel began to arrive. This is yet more proof of how quickly this operation was pursued. Weisgall testifies, "The islanders were overwhelmed by all the fanfare, geologists, botanists, biologists, and oceanographers categorized the flora and fauna of the atoll, and engineers blasted a deep-water channel through the reef to the beach on the main island of Bikini. Meanwhile, the Bikinians, who had never before seen motion pictures, were entertained with Mickey Mouse cartoons, Roy Rogers westerns and Hollywood bedroom farces."[66] This was an old-style colonial tactic: gain the trust of the host by paying a largely symbolic fee for the possibility of extracting something much more valuable. It is not accidental that in post–World War II military-scientific sites, the portable cinema apparatus[67] found yet another mission. In the Marshall Islands it participated in opening up ways for even more filmic production and the creation of the bomb archive.

Initially invented for the leisure of soldiers to keep up their spirits between battles, portable film projection was used to impress the islanders who were getting acquainted with both ends of "film culture"—viewership and acting—and the entire spectrum of exploitative tendencies in-between. It is obvious that performances took place without release forms and without parties involved fully understanding the script. Interpreting this material that reverberates through several iterations of productions necessitates a very broad analysis of the context. In this case, categories of fiction, documentary, and documentation are also blurred. I reiterate that the parameters shaping the conditions of film production and falsified circumstances of agreement to abandon the islands allow us to draw parallels with settler-colonial treaties.

The human subjects, with no access to authentic and personal (as opposed to scripted) verbal expression on screen, had to be inserted into a

---

66. Weisgall, "The Nuclear Nomads of Bikini."
67. Wasson, "Experimental Viewing Protocols."

fully scripted audiovisual narrative that was meant to hold evidence, act as a contract. Historically, an academic discipline of visual anthropology primed US military-artistic crews for such colonial audiovisual productions. If fabrication at the moment of dispossession from the Indigenous populations or fabrication for the sake of dispossession was not a new practice, its filmic nature was.

For US citizens, apart from the dominant "savage" tropes, newsreels provided what Priya Jaikumar calls "accurate imagination"[68]—a strategy applied by British educators in audiovisual travelogs teaching subjects located in Great Britain about remote colonies. Such emphasis on colonial visual education was proposed by a British imperial geographer, Halford Mackinder, author of the notion of geopolitics,[69] in order to foster a unified vision of the empire, the grounds for which were laid many years ago. The audiovisual tools were meant to strengthen ties between subjects of the colonial center and colony at a time when decolonial movements were picking up in the twentieth century. While narrating the "Cold War inevitable" and inventing post–World War II securitization discourse, the US newsreels from the Marshall Islands followed the logic of "accurate imagination."

To dethrone the usual framing of Pikinni Atoll as a site of the Cold War arms race, I want to view the Pikinni Atoll primarily as a film production set. In this sense, what it takes to produce the image, the visuality of the "special effect," is an inclusive part of image production. The Marshall Islands "tests" could be seen as a "runaway production." The Trusteeship Agreement delineated the borders of the set. Production costs were much "cheaper" outside the United States, and the end product became an "international success," which soon entered the cinema canon.

---

68. Priya Jaikumar, "An 'Accurate Imagination': Place, Map and Archive of Spatial Objects of Film History," in *Empire and Film*, ed. Lee Grieveson and Colin MacCabe (London: British Film Institute, 2011), 167–88.
69. Mackinder, "The Teaching of Geography from an Imperial Point of View, and the Use Which Could and Should Be Made of Visual Instruction," in *Empire and Film*.

Decolonial analysis of the bomb archive gives insight into how ocean space, assumed to be a vacuum, in conjunction with nuclear technology produces a nuclear colony with its ongoing nuclear and image-anchored violence. As part of this broader process, the roots of the nuclear colony in the twentieth century's anti-imperial context led to the particular type of audiovisual archive of newsreels and images of bomb explosions that are as abundant and available as they are misleading. I want to link this archive to manifestations of power, which Oldenziel describes as "often and purposefully . . . hidden from view."[70] In the analysis of US power deterritorialization, Oldenziel claims that "the U.S. wields a strikingly different kind of power because it lacks overseas possessions. . . . the U.S. does not occupy vast tracts of land outside the American continent like the Roman, British, and Russian empires of yore. But the U.S. does rule over extensive—but to its citizens, invisible—island possessions,"[71] which serve as technopolitical nodes. I emphasize that, in the case of the Marshall Islands, an oceanically deterritorialized technopolitical extraction is also enacted through the production and circulation of the image.

It is not the invention of nuclear technology per se, but the international legislation, backed by ethical and aesthetic paradigms, in combination with a persistence of the settler-colonial logic, that allowed direct and metaphorical atomization of the twentieth century's US colonial project. Nuclear weapons explosions for cameras are moments of violent alteration of the human relationship with the ocean in the vast areas in the Pacific, whose effects, despite all the hypervisibility they offered, have failed to translate into moral and political accountability. Following Oldenziel's description of the deterritorialization of power[72] as expansionism without juridically registrable and materially visible evidence, I claim that the bomb archive serves the function of audiovisual deterritorialization. Under such circumstances,

70. Oldenziel, "Islands," 13.
71. Oldenziel,14.
72. Oldenziel.

a film-industry analytical framework is most capable of recognizing and exposing such injustices.

Habitually, following the US Cold War logic, the Marshall Islands nuclear weapons explosions are framed as a demonstration of US power to the USSR and the creation of a "balance of power" in the world. However, representations of these explosions signify and perpetuate an invisibilizing of Indigenous experiences. I extend the notion of the bomb archive to the combination of representations of the nuclear weapons "tests" and their archiving practices. The processual nature of long-lasting nuclear weapons effects is embedded not only in the elemental violence of radioactive contamination but also in image production, distribution, and its archiving. In this sense, the image has been both a goal and an instrument in what is called "testing" of nuclear weapons. These images, produced at the time of nuclear weapons production, form a toxic archive that should be addressed through epistemic categories other than those relevant to the Cold War, which produced differential treatment of islanders.

The nuclear weapon "tests" in ocean space allowed the United States to "bracket" open waters, first legally as a no-go zone available for "tests," and later materially, as an excessively contaminated place. Of course, the effects of these operations—the contamination patterns—seem to have never perfectly aligned neither with juridical nor military-scientific "bracketing," which points at the broadest scope of ecological concerns.[73]

Conclusion

The US Trusteeship Agreement provided a framework for extraction without ownership that would entail not only having to shift the oceanic regime but also produce a different type of accountability for its citizens. The Indigenous population remained entirely "other," both symbolically, documented

---

73. Johnston, "Nuclear Disaster."

as "savages,"[74] and juridically, as non-US citizens. While acknowledging the ever-evolving complexities around changing modes of the Republic of the Marshall Islands and diasporic governing,[75] a focus on the juridical and material configuration in the broadest sense can help us understand why US military crews went to such lengths to produce a newsreel depicting an agreement with islanders in order to evacuate the Pikinni Atoll. This audiovisual document stands out for its function—treaty-like evidence at the moment of dispossession.

Through an elaborate production of the image of nuclear weapons "tests," the United States and the United Nations instigated a shift in the oceanic regime toward what Oldenziel calls a deterritorialization of US power through island occupation. The production of the image is inextricable in this instance. The US-UN Trusteeship Agreement allowed for the United States to cast onto the Marshall Islands a Euro-American conceptualization of ocean space as a vacuum. This settler-colonial configuration is produced through an elaborate combination of material, symbolic, and juridical factors, as this essay has attempted to sketch. Understanding the Pikinni Atoll as definable through settler-colonial dynamics allows us to place analytic emphasis on the extensive, ongoing nature of extraction and grasp the mediated core of it. To understand the dynamics at hand, it is necessary to depart from event-based narration and focus instead on process-based conceptualization. Ultimately, the United States, while on a mission of temporary

---

74. Davis, *The Empires' Edge.*
75. For more on challenges related to the national and international legislation intersecting in complex ways in the Marshall Islands, see Anita Smith, "Colonialism and the Bomb in the Pacific," in *A Fearsome Heritage: Diverse Legacies of the Cold War,* ed. Arthur John Schofield and Wayne Cocroft (London: Routledge, 2016), 51–71; Anita Smith and Cate Turk, "Customary Systems of Management and World Heritage in the Pacific Islands," in *Transcending the Culture-Nature Divide in Cultural Heritage,* 23–34; Martha Smith-Norris, *Domination and Resistance: The United States and the Marshall Islands during the Cold War* (Honolulu: University of Hawai'i Press, 2016); Schwartz, "Marshallese Cultural Diplomacy in Arkansas"; Schwartz, *Radiation Sounds*; Susanne Rust, "How the U.S. Betrayed the Marshall Islands, Kindling the Next Nuclear Disaster," *Los Angeles Times,* November 10, 2019; Davis, *The Empires' Edge.*

trusteeship, has turned several islands into a zone of near-permanent dam-age. The inconceivable longevity of radioactive contamination means that for all intents and purposes we are talking about a theft of ocean waters and islands.[76] The Trusteeship Agreement in combination with nuclear damage entirely bypasses the discoverist terra nullius sequencing of the settler col-onizer's gradual taking over of the land, which is often structured around divergent priority of rights to buy land versus Indigenous people's rights to retain or to sell it.[77]

Due to US nuclear weapons production, the Pikinni Atoll became less of a place that one can inhabit or visit and more of a mediascape. When compared to other instances of the application of audiovisual technology, the US mili-tary's takeover of the Pikinni Atoll is remarkable in that the ocean waters and islands were being inscribed into a settler-juridical order at a time when cin-ema production and circulation technology was already invented and widely available. This means that islanders appear on-screen not just to give testi-monies of colonial atrocities in a post-contamination landscape, as it appears in audiovisual representations of other (settler)colonial cases. The Marshallese community has a performed removal from ancestral waters and islands docu-mented on film. In this case, the cinema technology amalgamates an intense collection of temporalities: an early settler-colonial method is being applied at the moment of birth of nuclear modernity, during the time of decolonization.

Meanwhile the initial archive of raw footage made up of snippets and entire newsreels continues to circulate as US archival property, and when it is incorporated in other productions, through creative or scientific engage-ments, it becomes either corporate or private property. Such gradual increase of private/corporate ownership rights is comparable to differential distri-bution of land ownership rights in settler-colonial contexts, the juridical and ethical genealogies of which have been analyzed by Robert Nichols.[78]

---

76. Nichols, *Theft Is Property!*
77. Nichols.
78. Nichols.

I am suggesting here, therefore, that we understand settler coloniality and its violence not solely through national borderscapes but also through mediascapes.

There is neither information accompanying the sources correcting the facts presented in the newsreels nor information about atrocious production "costs" of the entire bomb archive. One could say that, legally, the government of the United States does not need consent to be able to use any type of bomb archive footage. However, exploring the possibilities for making this precedent at the very least deserves a discussion. Such a move would allow for the exploration of how intellectual property rights, and the audiovisual sphere in the broadest sense, could become a more literally and metaphorically visible and accessible site for articulating return and reparation demands. The bomb archive is part and parcel of the same data-classification paradigm as the health records of the Marshallese "test" subjects. Questioning the givenness of the right to screen nuclear colonial productions or creating a specialized fund collecting screening fees that would directly contribute to the impacted communities could help draw attention to the persistence of nuclear violence.

The issues caused by the classification, categorization, and canonization of audiovisual nuclear archives extend to the present and perpetuate *extraction through image*. In an effort to shift away from fetishizing both the bomb and the visual technology that in the Marshall Islands has been used as a strategy of deterritorialization, I propose incorporating an acknowledgment of worlds destroyed every time the bomb archive gets (audio)visually evoked. This could help distinguish between productions that are building on invisibilizing legacies of Cold War epistemologies and productions that are participating in the self-determination of communities that are enduring the deterritorializing violence of the mediascape-entrenched bomb archive. Such an approach each time would require making visible both the systemic violence of militarization and putting emphasis on different experiences implied in images that most often are articulated through notions of "American nuclear modernity." The forms of acknowledgment and definition of other engagements with nuclear archives should be produced by the impacted communities. My hope is that such a reclamation of agency in the area of representation could help include Marshallese experiences in established nuclear epistemologies, with space for dignity.

Acknowledgments

I would like to express my immense gratitude to my teachers Masha Salazkina, Krista Lynes, and Luca Caminati, as well as Lucas Freeman, guest editor of this issue Kenneth Paul Tan, and anonymous reviewers for generous feedback shared with me in different stages of research and writing. This article would not have been the same without my wonderful colleagues in the *Repair, Reparations, and Repatriation* study group, which was initiated and run by Jason Fox and Sameer Farooq. Our collective work in Winter-Spring 2021 engaging with the politics and practices of reparations and repatriation through the lens of a politics of repair, empowered me in my research and writing that among other things resulted in this article.

Das unsichtbare Visier—A 1970s Cold War Intelligence TV Series as a Fantasy of International and Intranational Empowerment; or, How East Germany Saved the World and West Germans Too

TARIK CYRIL AMAR

Abstract

This article addresses a franchise of intelligence films in the former Communist East Germany. Under the general title *Das unsichtbare Visier—The Invisible Visor*—they were produced for television and very popular. In general, the Cold War East produced a rich array of its own intelligence heroes, which cannot be reduced to mere derivatives of Western models. Yet there were commonalities and interactions across Cold War divides. One of these common features that *Visier* shared with many intelligence films globally was the depiction of abroad as both an "invisible front" of dangers and temptations and an exciting realm of adventure and consumption. *Visier* could not but also be a fantasy about East German citizens encountering, withstanding, and also enjoying the dangers and temptations of the Western Cold War Other. This included their facing two peculiar challenges: a degree of international mobility unlike anything the vast majority of ordinary East Germans could experience and the West's consumerism. *Visier* addressed both these issues through what we could describe as an essentially playful—and

https://doi.org/10.3998/gs.2489

dis-playful—practical cosmopolitanism. A careful look reveals *Visier* as a rich artifact of Cold War popular culture, with complex messages. The image of the heroic East German agent included a running comment of compensatory wish fulfillment. Here were ideal East German citizens doing their duty and yet also getting a fair slice of the capitalist good life abroad that most of their compatriots could not reach. They also consistently punched above their weight vicariously for East Germany as a whole. Like Britain's James Bond, these were agents of an at-best middling power doing major things in the world at large. And finally, perhaps most satisfyingly of all, they turned into gentle, benevolent guardian angels of hapless West German cousins, neatly reversing West Germany's claims of superiority.

Keywords: Cold War, Popular Culture, Intelligence Agents, Television, East Germany, James Bond

Introduction

During the last century's Cold War, narratives about intelligence agents (or, simply put, spy fiction) was a booming entertainment genre in both the West and East.[1] This article contributes to the history of Cold War popular culture—with "popular" understood in a modern, not folkloristic or premodern sense, and "culture" in a broad, not high-culture sense—by focusing on an example of spy fiction on television.[2]

---

1. *Cold War* was mostly a Western term and using it for its Eastern side should not "level the diverse experiences, mentalities, and practices connected to the forty-year standoff between the Eastern and the Western camp." See Anette Vowinckel, Marcus Payk, Thomas Lindenberger, "European Cold War Culture(s)? An Introduction," in *Cold War Cultures: Perspectives on Eastern and Western European Societies*, ed. Annette Vowinckel, Marcus M. Payk, Thomas Lindenberger (New York: Berghahn Books, 2012), 1.
2. My understanding of popular culture draws on multiple sources, including Adele Marie Barker, "The Culture Factory: Theorizing the Popular in the Old and New Russia," in *Consuming Russia: Popular Culture, Sex, and Society since Gorbachev*, ed. Barker (Durham, NC: Duke University Press, 1999); Lila Abu-Lughod, *Dramas of Nationhood: The Politics of Television in Egypt* (Chicago: University of Chicago Press, 2005); Jason Dittmer, *Popular Culture,*

Television has special relevance in this context because in the postwar period the rise of the fictional spy to hero status—a process that began in the nineteenth century but escalated in the second half of the twentieth century (and is continuing, for better or worse)—coincided with the emergence of

---

Geopolitics, and Identity (Lanham, MD: Rowman and Littlefield, 2010); Thomas Hecken, *Theorien der Populärkultur. Dreißig Positionen von Schiller bis zu den Cultural Studies* (Bielefeld: Transcript, 2007); and John Street, *Politics and Popular Culture* (Philadelphia: Temple University Press, 1997). Scholars from various disciplines have addressed the relationship between culture and the Cold War. Their contributions cannot be summarized or surveilled here. Important works include Stephen J. Whitfield, *The Culture of the Cold War*, 2nd ed. (Baltimore, MD: Johns Hopkins University Press, 1996); Frances Stonor Saunders, *The Cultural Cold War: The CIA and the World of Arts and Letters* (New York: W. W. Norton, 2000); David Caute, *The Dancer Defects: The Struggle for Cultural Supremacy during the Cold War* (Oxford: Oxford University Press, 2003); Annette Vowinckel, Marcus M. Payk, and Thomas Lindenberger, eds., *Cold War Cultures: Perspectives on Eastern and Western European Societies* (New York: Berghahn Books, 2012); Christopher Moran, "Ian Fleming and the Public Profile of the CIA," *Journal of Cold War Studies* 15 (2013), in particular, on the place of film in Cold War culture and cultural Cold War; Kirsten Roth-Ey, *Moscow Prime Time: How the Soviet Union Built the Media Empire that Lost the Cultural Cold War* (Ithaca, NY: Cornell University Press, 2011); Thomas Doherty, *Cold War, Cool Medium: Television, McCarthyism, and American Culture* (New York: Columbia University Press, 2003); Frederick Barghoorn, *The Soviet Cultural Offensive: The Role of Cultural Diplomacy in Soviet Foreign Policy* (Princeton, NJ: Princeton University Press, 1960); Kenneth Osgood, *Total Cold War: Eisenhower's Secret Propaganda Battle at Home and Abroad* (Lawrence: University of Kansas Press, 2006); Saunders, *The Cultural Cold War*"; Peter Coleman, *The Liberal Conspiracy: The Congress for Cultural Freedom and the Struggle for the Mind of Post-War Europe* (New York: Free Press, 1989); Paul Lashmar and James Oliver, *Britain's Secret Propaganda War 1948–1977* (Stroud: Sutton, 1998); Hugh Wilford, "Calling the Tune? The CIA, The British Left and the Cold War, 1945–1960," and Richard J. Aldrich, "Putting Culture into the Cold War: The Cultural Relations Department (CRD) and British Covert Information Warfare," both in *The Cultural Cold War in Western Europe*, ed. Krabbendam Scott-Smith (London: Routledge, 2004); Peter Busch, "The 'Vietnam Legion': West German Psychological Warfare against East German Propaganda in the 1960s," *Journal of Cold War Studies* 16 (2014); Michael David-Fox, "The Iron Curtain as Semipermeable Membrane: Origins and Demise of the Stalinist Superiority Complex," in *Cold War Crossings. International Travel and Exchange across the Soviet Bloc, 1940s–1960s*, ed. Patryk Babiracki and Kenyon Zimmer (College Station: Texas A&M University Press, 2014); Moran, "Ian Fleming and the Public Profile of the CIA"; and Lowell H. Schwartz, *Political Warfare against the Kremlin: US and British Propaganda Policy at the Beginning of the Cold War* (London: Palgrave Macmillan, 2009), 211. For a challenge to the consensus view of the Cold War's importance, stressing the limits of its influence on, in this case, American culture, see Peter J. Kuznick and James Gilbert, eds., *Rethinking Cold War Culture* (Washington, DC: Smithsonian Institution Press, 2001).

traditional (predigital) television as a mass medium. The result of this conjuncture was that while we may associate spies as popular-culture icons with characters such as James Bond or Jason Bourne, who started out on book covers but who have achieved peak fame via cinema, in reality, television has played an at least equally important role in bringing spy fiction to mass audiences.

Indeed, if we turn our attention specifically to the East of the Cold War (here meaning the Soviet Union and its allies/clients in eastern and central Europe), we find that in three countries at least the single most popular piece of spy fiction was made for television: In Communist Poland and the Soviet Union the blockbuster series "*Stawka większa niż życie* (Stakes Greater than Life) and *Semnadtsat Mgnovenii Vesny* (Seventeen Moments of Spring), respectively;[3] and in East Germany, a kind of franchise of spy thrillers under the title *Das unsichtbare Visier* (The Invisible *Visor*—hereafter *Visier*), shown between 1973 and 1979.[4] This article will focus on the later instalments of *Visier* after 1976.[5]

---

3. *Semnadtsat Mgnovenii Vesny* ("Seventeen moments of spring") was based on a novel, which formed part of Iulian Semionov's series of stories about its hero, a fictitious Soviet agent usually known by his German cover name Shtirlits (or Max Otto von Stierlitz). Yet the Shtirlits that became a persistent pop-culture icon in Russia (and not only) is the one produced by Semionov (as [co]script writer), the director Tatiana Lioznova (also de facto a coscript writer), and, last but not least, the brilliant actor Viacheslav Tikhonov, functioning in essence as Shtirlits's Sean Connery. In the case of *Stawka* and *Visier*, the films came first; stories and novels were spin-offs.

4. The period 1973 to 1979 refers to the first showings of the films. At the same time, reruns of earlier parts of *Visier* began as early as 1974 and did not end until after 1979. *Visier*'s structure, discussed in more detail later, is not easy to categorize. In official usage it was called, for instance, *eine mehrteilige Filmerzählung* (a film narrative in multiple parts), archive of the "Bundesbeauftragter für die Unterlagen des Staatssicherheitsdienstes der ehemaligen Deutschen Demokratischen Republik (hereafter BStU) MfS ZAIG 26967: 20. In English, the concept of the franchise comes reasonably close. At least sometimes, commentators also used the term "Kundschafterserie" (as, for instance, "Depot im Skagerrak," *Volksarmee*, no. 36, 1977, which means a series about *Kundschafter*, a euphemism for spy, discussed further.

5. In all three countries—Poland, the Soviet Union, and East Germany—there were other, sometimes very successful spy stories in the shape of novels and films (for cinema and television). But in each country, we can clearly identify one case of outstanding popularity; namely, *Stawka*, *Vesny*, and *Visier*, respectively.

Das unsichtbare Visier (The Invisible Visor), Phase 2

A separate discussion of only this second phase makes sense because the whole of *Visier* consisted of two related but substantially different parts: the first nine films, shown between 1973 and 1976, featured a single hero agent called Achim Detjen (his cover identity), played by the East German (and later international) star Armin Mueller-Stahl. [6] After his departure (due to a combination of his own dissatisfaction and political disfavor), a second set of seven films, released between 1977 and 1979 under the same title, but in fact quite different, showed a team of agents.

While this article is about the second, post-Mueller-Stahl, team-based version, it occasionally refers to the first, single-hero iteration, too, as well as to *Visier* as a whole, encompassing its first and second phase. I have adopted a shorthand, which, readers should keep in mind, has not been used officially or generally; it is merely a device for making this text easier to read: When referring to the single-agent films turning on Achim Detjen's adventures alone—*Visier* I; for only the second, team-based version—*Visier II*; and for *Visier* I and II together—simply *Visier*.

---

6. Mueller-Stahl did not "defect," and *Visier* did not end after he stopped playing the main character (as sometimes stated in the literature). Beyond the team version of the franchise, there was also a later spin-off, "Fire Dragon" (Feuerdrachen) and an unrealized plan for a second spin-off under the title "Jungle of Missiles" (Raketendschungel). See Sebastian Haller, "Imaginations of Insecurity: Representations of the State Security Service in East German Television in the Late 1960s and 1970s," in *Socialist Imaginations Utopias, Myths, and the Masses*, ed. Jakub Beneš, Stefan Arvidsson, and Anja Kirsch (London: Routledge, 2018), 206. The first films, with Mueller-Stahl as key protagonist, are the subject of my discussion of *Visier* in a book to come out soon. On *Visier* and other East German intelligence films, see also Haller, "Imaginations of Insecurity"; Haller, "'Diesem Film liegen Tatsachen zugrunde . . .': The Narrative of Antifascism and Its Appropriation in the East German Espionage Series 'Das unsichtbare Visier (1973–1979),'" *History of Communism in Europe* 5 (2014); Carol Anne Costabile-Heming, "Espionage and the Cold War in DEFA Films: Double Agents in *For Eyes Only* and *Chiffriert an Chef—Ausfall Nr. 5*," in *Cold War Spy Stories from Eastern Europe*, ed. Valentina Glajar, Alison Lewis, and Corina l. Petrescu (Lincoln: University of Nebraska Press, 2019).

Visier, it should be made explicit, has no claim as a work of film art. At its best, it was solidly made television entertainment with a political message, which was sometimes fairly unobtrusive and implicit (especially in *Visier I*) and sometimes crudely blunt and preachy (in particular in *Visier II*) but always pervasive. Like most films about secret agents everywhere, *Visier* was anything but realistic.

Yet in a franchise that lasted seven years and featured sixteen individual films (which usually formed clusters of two or three with one story arc), there were bound to be differences. In that respect, the films of *Visier I* clearly showed better craftsmanship, while those of *Visier II* marked a conspicuous decline of the series. This was not a matter of the acting: while *Visier I* had an extraordinarily gifted and even charismatic actor playing its main protagonist, those taking over in *Visier II* were competent professionals as well. It is true that Mueller-Stahl deliberately shaped the role of Detjen as he saw fit; namely, as an "adventurer" rather than a Communist "party man," as he later put it. This input, too, was lost when he left. But that was not the key reason for *Visier*'s subsequent deterioration.

Rather, it was the writing that changed, and we know how: the scripts for *Visier I* had been produced by a pair of experienced authors, Otto Bonhoff and Herbert Schauer. While deeply politically conformist, whether out of conviction or opportunism or both, and thoroughly middle brow in terms of literary skills, they could be relied on to put together a plausibly solid plot within the conventions of the genre. But, while Bonhoff and Schauer had worked closely with the Stasi's Office of Public Relations—in German the "Abteilung Agitation," "Pressestelle," or "Presseabteilung"—after 1976, the writing was taken over by the head of the office itself, Günter Halle, a Stasi colonel with literary ambitions who from then on appeared under the pseudonym Michel Mansfeld in *Visier*'s credits.

Halle, however, was clearly out of his depth: the three stories—in seven films—produced during this period are shot through with glaring inconsistencies, comically clunky twists and turns, and basic errors, such as pedestrian expositions. Even making allowance for the conventions of the genre and the needs of popular entertainment, they were exceedingly shoddy. Indeed, Halle's scripts were so obviously inept that *Visier II* found harsh critics who openly denounced its convoluted and confusing plots and, in one case, at least, went so far as to ask if the series had lost its drive.[7]

Put differently, if *Visier I* was efficient 1970s popular entertainment that often looks a little quaint only now, in retrospect, *Visier II* was really quite bad by any standards, simply in terms of scripting craft. Yet this study is not concerned with artistic quality but political meanings.

The East German *Kundschafter* Idealized: Cosmopolitan and Decisive

Regarding those meanings, what all *Visier* films—*I* and *II*—had, unsurprisingly, in common was their highly idealized depiction of the agents of East Germany's combined secret police and intelligence service, the Ministry of State Security (in short, Stasi), referred to, when working in foreign intelligence at least, by the euphemistic term *Kundschafter*, literally meaning a scout or reconnaissance soldier. A calque from the Russian *razvedchik*—used in the same manner by and| about the Soviet secret services, the KGB and GRU—the word *Kundschafter* also signaled the Stasi's and East Germany's alignment with the Soviet Union. The fact that, in reality, the preponderant majority of Stasi personnel (about 90 percent at least) were not engaged in foreign espionage but domestic surveillance and repression found no

---

7. See "Neue Kundschafter im Visier," (New Kundschafter in the visor), *Der Morgen*, December 19, 1977. By 1981, Halle/Mansfeld's incompetence—and perhaps a failing intuition for the latest political trends as well—led him to fail even more obviously: his "Feuerdrachen"—a stand-alone spy thriller originally planned as a *Visier* sequel—was panned by critics as confused and wooden. See Peter Hoff, "Aufhellung eines dunklen, gefährlichen Geschäftes: 'Feuerdrachen,' ein Film des Fernsehens der DDR" (Illumination of a dark, dangerous business: 'Feuerdrachen,' a film on GDR television), *Neues Deutschland*, December 24, 1981, 4.

reflection in *Visier*.[8] In effect, a small minority of agents—those spying abroad—were highlighted in order to cast an aura of patriotism, adventure, and courage over the many who were, by any standards, home-front enforcers of an authoritarian regime.

As mentioned before, while the Soviet context was specific to the eastern side of the Cold War, *Visier* was also part of a global (East and West, and beyond as well) phenomenon of Cold War popular culture—namely, the systematic heroizing and popularizing of (usually fictitious) intelligence agents. In the West, and then globally, the single most extreme and well-known case has been James Bond, a character invented by middle-brow writer Ian Fleming that, from the early 1960s, became the basis for one of the most profitable movie franchises in the history of cinema.

Yet, as noted above, while often overlooked, the Cold War East produced a rich array of its own intelligence heroes, which cannot be reduced to mere "answers to Bond" or derivatives of Western models in general. At the same time, it is true that, apart from differences, there were commonalities and interactions across Cold War divides. One of the common features that *Visier* shared with many intelligence films globally was the depiction of abroad as both an "invisible front" of dangers and temptations and an exciting realm of adventure as well as consumption. This was a feature that *Visier* clearly shared with the global genre: the well-traveled secret agent as (also) a well-off and suave consumer was a fantasy prominent in Western Cold War culture too.[9] But, of course, in *Visier*, anything linked to

---

8. Jens Gieseke, "East German Espionage in the Era of Détente," *Journal of Strategic Studies* 31 (2008): 400. In 1989, the Stasi had ninety-one thousand employees, with nine to ten thousand working on foreign intelligence (plus another eleven hundred in military intelligence). These figures were the result of significant expansion: Stasi foreign intelligence in its main branch (the HVAHauptverwaltung Aufklärung) had counted about five hundred to six hundred staff members in 1958, almost three thousand in 1982, and over forty-seven hundred in 1989 (with the same again in various other Stasi branches outside the HVA). See Gieseke, "East German Espionage in the Era of Détente."

9. On the spy and cosmopolitan fantasies, for instance, see Craig Calhoun, "Cosmopolitanism in the Modern Social Imaginary," *Daedalus* 137 (2008): 106.

mobility across borders and consumption acquired special—though not explicit—associations and significance because East Germany was highly restrictive about letting its citizens travel, and the West in general and West Germany in particular represented the temptation of a conventionally superior, capitalist level of consumerism.

In this regard, it is important to note a peculiarity of the East German situation (and, literally, location): a twentieth-century authoritarian-socialist state with advanced if largely comparatively inefficient industry and typical problems satisfying consumer desires, it was also constantly exposed to West German television for most of its existence. Almost everywhere in their country, East Germans were tuning in not only to news from the capitalist Cold War opponent (who happened to speak the same language) but shows, films, and advertisements displaying, often in idealized form, the consumer advantages of the West.

Thus, *Visier* could not but also be a fantasy about exemplary East German citizens encountering, withstanding, and also enjoying the dangers and temptations of the Western Cold War Other. This included their facing two specific—if always implicit—challenges: a degree of international mobility that was entirely unlike anything the vast majority of ordinary East Germans could experience and the West's advanced consumerism. *Visier*, in effect, addressed both of these issues through what we could describe as an essentially playful—and literally dis-playful—practical cosmopolitanism. Its heroes' sartorial style ranged from elegantly casual to full evening dress. They lived in well-appointed modern apartments and houses, which featured comforts such as pool tables. Viewers saw them hobnobbing with the unsuspecting Western establishment at exclusive and opulent parties, in fine-dining restaurants, occasionally even in a nightclub. They stay at pleasant hotels where they sometimes do their laps in beautiful pools and they even talk openly about wanting to go shopping while in Rome. In *Visier II*, one of the *Kundschafter* team drives an elegant Western vintage sports car and takes Bondesque bubble baths with his girlfriend.

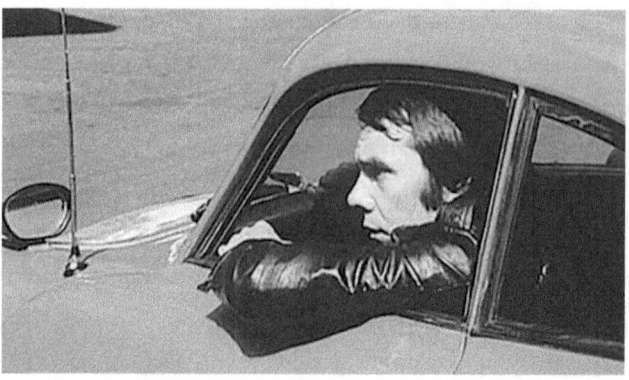

Figure 1: Stasi Kundschafter Tanner in a red vintage Porsche.
© Fernsehserien.de.
Source: © Fernsehserien.de

Examples could be multiplied but the key point is clear enough: it would be very misleading to imagine *Visier*'s East German agents, from a regime committed to Communism, as hiding among the Western proletariat or agitating workers. On the contrary, they were depicted as assuming elite identities, such as staff officers, politically and socially well-connected lawyers, or hipsterish fashion photographers. This, of course, made dramatic sense: all spies, after all, have to adapt to their target by dissimulation. Yet, at least in terms of effects, there was more to this facet of *Visier* than the necessities of the plot. It entailed an elite group of, by definition, exemplary East German citizens enacting the dream of having it both ways: being committed in perfect loyalty to East Germany and its official values, ideology, and careers while, at the same time, not paying the usual price of (relative) consumer deprivation.

At the same time, in the case of the Cold War midsize power East Germany, its agents' adventures beyond its borders also served as a fantasy of international reach and influence. Stories about victories in an area that was, by definition, opaque and thus especially open to making things up were ideal instruments of psychological compensation: it is in the nature of victories on an invisible front that they cannot be verified.

Of course, despite its specific East German conditions, in principle this compensation fantasy effect, too, was anything but unique, and we could find it in other national contexts as well: James Bond has embodied the same escapism from a reality of postwar decline. Despite Great Britain's special relationship with the United States, especially after the independence of India in 1947 and the debacle of Suez in 1956, post–World War Two Britain has been a former imperial power that sometimes finds it difficult to adjust to its diminished influence.[10] In a Cold War context, it did have more autonomy and clout than East Germany, exemplified by its own nuclear forces, but not enough to change the fact that it was dependent on its hegemon, the United States.

In Bond movies, this dependency was recast as a partnership, tilted, moreover, in Britain's favor: while the terms of their cooperation change over time, as Lisa Funnell has shown, Bond regularly works with Americans, draws on their assistance, makes friends among them, avenges them, and goes to bed with them as well. But Bond is always not only the "primary hero" but the central and by far most important figure.[11] When the stakes are highest, it is also the British super-agent, not the United States, who is truly indispensable.

A Special Case of Intranational Internationalism: East Germans Saving West Germans

There was yet another special—and crucial—aspect to *Visier's* relationship with the West: while the films, especially of *Visier II*, repeatedly imply that

---

10. Robbie B. H. Goh, "Peter O'Donnell, Race Relations and National Identity: The Dynamics of Representation in 1960s and 1970s Britain," *Journal of Popular Culture* 32 (1999): 31.
11. Lisa Funnell and Klaus Dodds, "The Anglo-American Connection: Examining the Intersection of Nationality with Class, Gender, and Race in the James Bond Films," *Journal of American Culture* 38 (2015): 371.

the East German Kundschafter save the West from itself—that is, the reckless and wicked schemes of its own official as well as concealed elites—this West consists of two categories of places, and one is more important than the other.

It is true that when, for instance, in a set of episodes under the title "The King Kong Flu," the Stasi agents battle CIA schemes to weaponize psychotropic drugs, viewers are given to understand that the Agency's evil experiments affect innocent Americans and perhaps even most of all. Thus, by striking a blow against them, *Visier*'s heroes are depicted as protecting these Americans as well, if indirectly. But they do so in West Germany, and the direct beneficiaries of their actions are West Germans. It was this constellation that was typical for *Visier II*. Its single most important motif was not, strictly speaking, international but intranational. Reflecting the special position of the two rival Cold War Germanies, it was a fantasy in which East Germans rescued West Germans.

Likewise, in *Visier*, the broader fantasy of international empowerment took a specific shape: Substantial parts of the stories unfold in countries

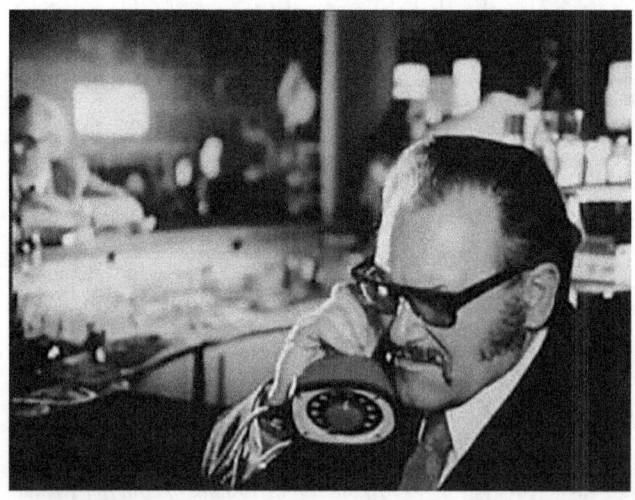

Figure 2: CIA Agent Wilson. © IMDb.
Source: © IMDb

that are unambiguously abroad (in the sense of not being Germany) and locations that carry associations of cultural difference and global interests (Argentina, Portugal, Norway, South Africa, Italy, France, Spain, and Corsica). Yet West Germany is at least equally important.

Moreover, whenever *Visier*'s agents go beyond West Germany, their operations are still linked to it. They end up in South Africa, for instance, in hot pursuit of surreptitious West German attempts to acquire nuclear weapons or in Italy and France on the tracks of a Western false-flag terrorist organization that involves the CIA and Italian conspirators as well as the West German military. Put differently, while all of the action takes place abroad (i.e., outside East Germany), *Visier*'s kind of being abroad often means living in West Germany, a country that is politically very different and geopolitically-ideologically located on the other side of the Cold War but, at the same time, another version of a shared national domain—if in a very contested manner.

This constellation had several implications and effects. At the most pedestrian level, it made it easier to tell stories about successful imitation and infiltration. Cold War East and West Germans shared a language, habits, and cultural codes—by no means perfectly but to an extent that made pretending to be the Cold War Other especially easy.

With regard to the politics of state legitimation, however, this close, special relationship touched on at least two extremely sensitive issues. For one thing, West and East Germans also shared a past, in particular that of Nazism and World War II. As is well known, in East Germany this fact was subject to a complicated form of denial. In essence, the East German regime absurdly defined itself and its society as derived only from the victims of and resisters against Nazism and projected the latter's legacies, ongoing aftereffects, and dangers on postwar West Germany alone. This position was one of the core tenets of East Germany's official doctrine of antifascism, which was second to none in legitimizing the country's existence as a separate and, so the claim, better Germany.

This aspect of East German ideology was central to *Visier I* in particular. In these first *Visier* films, released between 1973 and 1976 and set in the period between 1949/50 and 1961, the single hero Kundschafter's (Mueller-Stahl as Detjen) initial task was to fight an international network of unrepentant Nazis and militarists who are planning a comeback while successfully infiltrating West Germany's political, military, and intelligence elites. Even though this mission also involved contemporary issues, such as West German rearmament or alleged plans to provoke a war against East Germany, the background of Nazism—and Nazis personally—still alive and kicking, if covertly, was always present.

In *Visier II*, this theme is not entirely absent: a key villain is still a character introduced in *Visier I* as a former SS officer and war criminal who has eluded detection and punishment to turn into a wily postwar Nazi network cadre, well connected to crime, business, and intelligence. But the weight of this motif is clearly reduced. Instead, in *Visier II*, set in the 1960s, contemporary aspects dominate in the operations of its team of agents. Even when they face off against neofascists committing terrorist attacks under a Maoist false flag, the emphasis is on their opponents being precisely *new* fascists, often fairly young men and women produced in the postwar period, not merely old diehards of Nazism.

Another striking difference between *Visier I* and *II* has to do with how the films' Stasi heroes relate to ordinary West Germans, those who are not part of either Nazi or post-Nazi elites. In *Visier I*, such characters appear, but, mostly, they do not occupy much of the hero's attention. This was perfectly reconcilable with official East German thinking, which postulated explicitly that, ideally, ordinary citizens of capitalist states should manifest a "progressive" attitude by helping the Kundschafter from the "world socialist system."[12] Far from treason or spying, such assistance was to be understood as an "integral part of the international class struggle," according to, for

---

12. Tuchel, "Das Ministerium für Staatssicherheit und die 'Rote Kapelle'" (The ministry for state security and the 'red chapel"), 245.

instance, Karl-Heinz Biernat, a scholar and doctrine intellectual based at the East German Communist Party's Institute for Marxism-Leninism, a Communist think tank.[13]

In the same spirit, the operations of the Kundschafter were idealized as serving not simply the interests or even security of East Germany but the preservation of world peace as such, by helping the "socialist states under the leadership of the Soviet Union" defang the aggressive policies and militarist plots of the West.[14] In other words, the Kundschafter protected the ordinary citizens of the West as well—namely, from their wicked rulers. Complimentarily, these noble fantasy agents never targeted "the working people" (*werktätiges Volk*).[15] This was the meaning of the "humanism" for which *Visier* was often praised:[16] *Neues Deutschland*, East Germany's main newspaper, celebrated the films as combining thrilling entertainment, the "political unmasking" of Western interests and regimes, and a "great humane concern."[17]

There was, thus, nothing surprising about the fact that the hero of *Visier I* did no harm to ordinary West Germans or, indeed, ordinary people anywhere. On the contrary, he was repeatedly shown as getting along especially

---

13. Bundesarchiv (BA) DY 30/IV A 2/9.07/53: [380] ["DY 30/IV A 2/9.07/53: 380" is an archival signature.] After a hostile takeover of the Social-Democratic Party in 1946, the Communist party in East Germany called itself the Socialist Unity Party of Germany (SED).

14. Deutsches Rundfunkarchiv (DRA) Schriftgut, Bestand FS, Dramatische Kunst, Das Unsichtbare Visier, A 081–05–02/0001 TSig 001–010, "Herbert Schauer, creator, *Das unsichtbare Visier: Inhaltsangabe des zweiten Komplexes*," 3. East Germany, or authoritarian-socialist regimes in general, had no monopoly on the myth of espionage as securing peace. Introducing, for instance, a 1937 radio drama version of the 1915 spy thriller classic "The 39 Steps" (also made into a Hitchcock movie in 1937), an American military intelligence officer praised secret agents as "one of the greatest forces for world peace."

15. BStU MfS ZAIG 26967: 21. For an example of the special stress on saving the peace for West Germans as well, see "Wertvoller Gegenwartsfilm der DEFA" (A valuable contemporary film by DEFA), *Neues Deutschland*, July 21, 1963.

16. BStU MfS ZAIG 26967: 21.

17. "Abenteuerfilm 'Unsichtbares Visier,'" (Adventure film "Invisible Visor"), *Neues Deutschland (Berliner Ausgabe)*, December 27, 1973; and "Mörderischer Alltag im *unsichtbaren Visier*" (Murderous everyday life in *Invisible Visor*), *Neues Deutschland*, December 20, 1978.

well with them—for instance, when making friends with—and getting important information from—local farmhands and gauchos on a ranch in Argentina used as a Nazi hideout.

Visier II, however, went significantly further: its team of agents does more than merely abstain from hurting innocent Western bystanders or even protect them in the abstract, as indirect beneficiaries of that "world peace" guarded by the Kundschafter operations. In *Visier II*, they repeatedly rescue—or at least try to do so—individual West Germans, victimized by the dirty politics, flawed family and social relationships, moral hypocrisy, dark conspiracies, and ruthless secret service operations of the capitalist West, as shown and/or caricatured in *Visier*. Put differently, in these later films of the franchise, there is a clear new pattern, a motif of East Germans saving West Germans. It is crucial to note that while doing so the *Visier II* heroes even clearly go beyond—and take more risks than—what is necessary for their intelligence missions. Put differently, the Kundschafter are depicted as so idealistic and morally invested that they are more than good agents who also happen to be good human beings: in effect, they are recast as moralists who happen to be on an intelligence mission. And in that process, they de facto privilege the other Cold War Germans as in need of their protection.

"The Afrikaanse Broederbond": (Almost) Saving a West German Whistleblower

Thus, in the three films that together form the first story arc of *Visier II*, under the title "The Afrikaanse Broederbond," the team of Stasi agents battles against a complicated West German–South African ploy to transfer West German technology to the then apartheid regime in South Africa and build nuclear weapons together. Things get more complicated again, when it turns out that the American CIA is involved as well, seeking to control West Germany in the interests of US hegemony (if not necessarily to prevent West Germany's acquisition of the bomb).

At the same time, viewers learn that, at a research center in South Africa, Bonn (former West Germany's capital) and Pretoria are already cooperating on the development of chemical weapons of mass destruction (WMD). This part of the plot is essential, driving much of the films' action by focusing on the desperate struggle of a West German engineer, Jürgen Machholz, who works at the WMD development facility and contaminates himself with a new chemical agent by accident, which leaves him terminally ill. While his employers try, in essence, to murder him as now dispensable and inconvenient, he escapes to Germany where he hopes to find treatment. To facilitate the latter, he pilfers a sample of the secret as well as illegal substance that is slowly killing him. He also tries to make contact with a journalist. Put differently, although Machholz is no natural born rebel, his despair is turning him into a potential whistleblower. Clearly afraid of politically disruptive revelations, South African and West German agents try to find and kill him. When they do not succeed, they start targeting his wife.

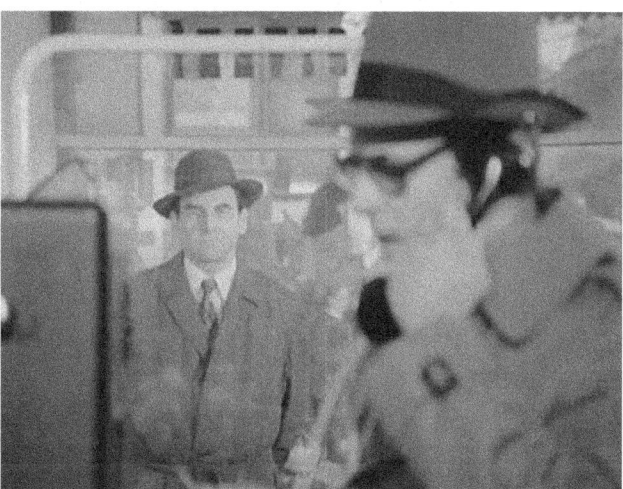

Figure 3: West German engineer and victim of capitalist intrigue Machholz looking for help, while *Visier* team leader Clemens keeps a watchful eye on him. © IMDb.
Source: © IMDb

Against this background, *Visier*'s team of Stasi agents not only under-
mines this nefarious West German–South African cooperation but also
dupes the CIA. They also save Machholz from an assassination attempt.
And while they cannot prevent his death from his initial contamination,
they manage to protect his wife and rescue her in a dramatic—as well as
perfectly implausible—operation in South Africa. The last thing viewers see
of her is how she escapes on a flight to Warsaw, clearly implying that she
will not return to West Germany but start a new life on the other side of the
Cold War divide.

King Kong Flu: Saving West Germans from the CIA

In the second adventure of *Visier II*, which unfolded in two films under the
title "King Kong Grippe" (King Kong Flu), chemical weapons play a key
role again. Only this time, it is the CIA itself that—in a clear if freewheeling
allusion to real, historic CIA research, such as Project MKUltra—is depicted
as running experiments with them, not only among unwitting citizens at
home but abroad as well, in this case in West Germany.

While not bereft of references to reality, the "King Kong Flu" episode
was still especially implausible on the whole and in detail: viewers were told
that experiments with weaponized drugs were conducted widely in American public space, in such a manner that random citizens could be exposed
while using an airport, the New York City subway, or a highway tunnel.
Moreover, the plot of "King Kong Flu" turned on the CIA running, or trying to run, the same kind of experiments on soldiers of the West German
army, a whole small town, and an individual provincial businessman, Alois
Leutwiler. The latter is systematically driven to suicide by a combination of
surreptitious drugging and devastating public revelations of his commercial
and political corruption as well as less than perfect family life—all for purely
experimental purposes.

If the story was far-fetched in general, its twists and turns were even more unbelievable: a secret West German army research facility (hidden under a medieval castle), also busy with chemical warfare, simply happens to be in the same nondescript town (seemingly) selected at random by a CIA computer for testing; Leutwiler also is an old friend of the CIA operative who directs his cumbersome assassination by psychological pressure and drugs; and, of course, that friendship is less than reliable, in part because the businessman's only daughter is, in reality, the CIA officer's daughter.

All of the above (and more) made for an especially scurrilous yarn that featured West (not East) German and CIA agents killing each other in and around an empty church in the West German back-of-beyond, a masked, ninja-like character scaling things in a black onesie, and a grand-finale explosion of dynamite conveniently left lying around in a hole in the ground since, at least, the postwar years yet still good to go.

Visier's heroes, meanwhile, secure a sample of not only the CIA drug but of the antidote as well, pointedly unlike the agents of the West German competition who are trying but failing to do the same. Clearly, with both drug and antidote finally in safe hands with the Stasi, viewers were meant to feel relief: surely, these samples would be put to purely defensive uses alone.

Yet, as in the preceding "Afrikaanse Broederbond" films, the Stasi Kundschafter manage to do even more—namely, recue West Germans, in this case en masse and individually. Regarding en masse, they foil the CIA's attempt to contaminate the local church's holy water. With respect to individuals, it is true that Alois Leutwiler cannot be saved from the American plot. But he is also a deeply unsympathetic character, a caricature of a rich, exploitative, and heartless capitalist who is also a corrupt politician from the CDU (West Germany's party of mainstream conservatism) and a criminal seeking to cover up a massive real estate scheme.[18]

---

18. In a very disturbing visual twist, one scene in Leutwiler's villa —in the far background— includes a candle holder that looks like a Jewish menorah. Otherwise, there do not seem to be any hints implying that Leutwiler is Jewish. The striking, if short, appearance of the menorah is hard to interpret: Was it meant as a deliberate if coded signal and thus an

Leutwiler's daughter, Georgia, really fathered by his false friend from the CIA, however, is an entirely different matter: an attractive young woman, she is depicted as altruistic and compassionate, if naïve. Viewers are clearly supposed to see her as an innocent victim of her family's hypocrisy. Put differently, she may have had the misfortune of being born into the rich bourgeoisie, but she is also a damsel in distress as well as a—local-scale—upper-class princess with a heart of gold. Her dad may be an ogre, his stately villa a cursed castle, but she needs not class struggle but rescuing. Which is what the youngest member of the Stasi team, the dashing "Genosse Alexander," masquerading as a fashion photographer with a serious art education, will do in the end. In a scene somewhat reminiscent of the American film classic *The Graduate*, "King Kong Flu" closes with Georgia leaving her corrupt philistine home for good, driving away together with Alexander.

And lest viewers misunderstand the meaning of this striking scene, a preceding conversation between Alexander and the leader of the Kundschafter

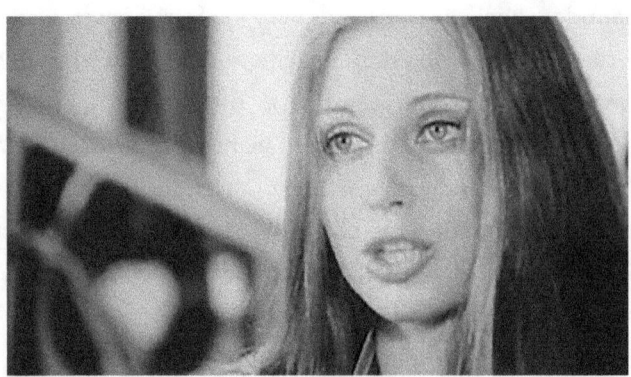

Figure 4: Georgia Leutwiler. © Google Play.
Source: © Google Play

instance of *Visier* conveying an anti-Semitic stereotype? Was it the outcome of sheer ignorance on the part of a scene decorator? Or perhaps meant as a nasty "joke" of some kind? From the available evidence, it is impossible to tell. But it is a detail that should be noted.

team, the older, kind, and distinguished "Dr. Clemens"—a cultivated man who knows his wines—makes clear two important points: Alexander genuinely cares for Georgia, notwithstanding that their putative random meeting was, in reality, initially merely a piece of spy tradecraft to get closer to the Leutwiler family.

Second, it is not only the young Alexander who believes that helping Georgia is an obvious priority, even if it has nothing to do with the intelligence mission of the team, which has already been completed. Rather, Clemens explicitly gives his blessing as well. In other words, Alexander picking up Georgia to help her escape her capitalist origins and confines is neither a James Bond-like final scene of sexual conquest, nor is it in conflict with the Stasi agent's orders. Those come from the sage Clemens, and his agreement signals that Alexander's rescue of Georgia is in alignment with the Kundschafter team's code of conduct. In sum, Georgia is saved not simply by a young man acting out of a mixture of romantic and moral motives who happens to also be a Stasi agent but by the Stasi itself: a Stasi that *Visier II* depicts as at least as keen to help the innocent victims of the West German political and social order as on reconnoitering its secrets. In essence, it is almost as if this motif of *Visier II* foreshadowed the American 1980s series *The Equalizer*, in which a retired CIA agent has nothing better to do than offer his help to whoever has the "odds against" them.

Island of Death: Preparing to Hand on the Baton?

Finally, there was "Insel des Todes" (Island of Death), the third adventure of the team of agents at the center of *Visier II*, and the last in the *Visier* franchise as a whole. Its two films made use of historic events of terrorism and political instability in postwar Western Europe—in particular the so-called Years of Lead in Italy—to tell a story of a secret campaign aiming at regime change in favor of the far right by a "strategy of tension": *Visier*'s heroes this

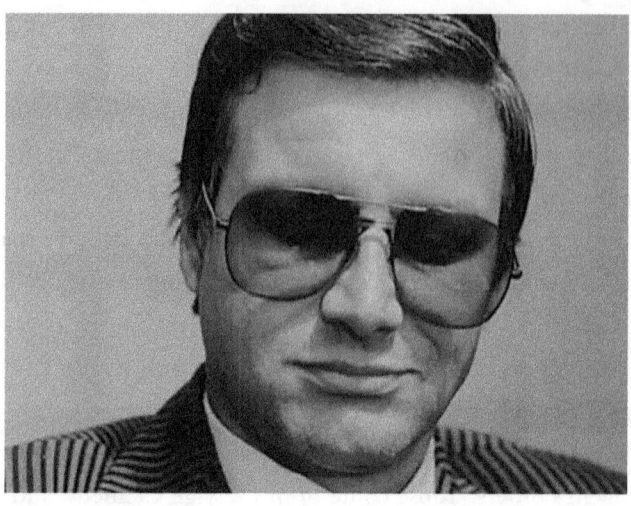

Figure 5: Moffo, the face of neofascist conspiracy in Italy. © IMDb.
Source: © IMDb

time have to battle a group of conspirators seeking to prepare the ground for several coup d'états, in Turkey, Greece, and Italy. Led by the CIA, these villains include various European government officials, neofascists, a few old Nazis, the West German secret service and army, and even some clueless Maoist and Trotskyist extremists. Staging false-flag terrorist attacks, the conspirators seek to implicate the Left and create opportunities for authoritarian takeovers from the Right.

Against this background, *Visier's* Kundschafter team manages to obtain revealing information about the conspiracy. Yet what makes "Insel des Todes" unique in the *Visier* franchise is that their activities are not dominating the action and often recede into the background. Instead, much of the center and foreground of the story are occupied by a character who has nothing to do with East Germany or the Stasi, a French everyman who turns into an unrelenting—and incredibly efficient—avenger when the neofascist network kills half his family.

One of the false-flag terrorist attacks consists of planting a bomb at the train station of the Italian city of Milan. Among the random victims are the wife and son of Marcel Laffitte.[19] Laffitte, who happens to have been a French paratrooper and explosives expert, sets out to find their killers and those behind them, quickly hitting on the regime-change campaign led by the CIA. As to be expected, after the requisite twists and turns, he succeeds: after infiltrating the conspiracy, he kills one of the two immediate perpetrators of the Milan bombing as well as the midranking cadre who ordered it and mines the plotters' high-tech headquarters—including a hidden submarine hangar—on an island off the French coast in the Mediterranean, the eponymous Island of Death.

Laffitte, it bears emphasis, unlike the Machholz couple from the "Afrikaanse Broederbond" or Georgia from "King Kong Flu," hardly needs any help. On the contrary, it is Laffitte who twice saves a member of the Kundschafter team, Winnie Winkelmann, not vice versa. The first time, when Winkelmann tries to reconnoiter the Island of Death and is almost caught,

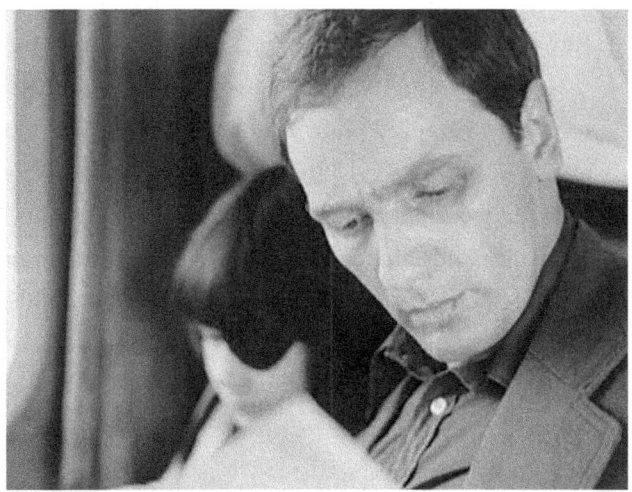

Figure 6: Marcel Laffitte and his surviving daughter. © IMDb.
Source: © IMDb

19. In an entirely different context, the name "Laffitte" occurs in Karl Marx's classic *The Class Struggles in France*. Its use in *Visier* may have been an ideological in-joke of sorts.

she gets away only thanks to Laffitte. Later, when an assassin tries to kill her, Laffitte saves her again.

Subsequently, Winkelmann and Laffitte cooperate to sabotage the neofascist network, crippling its arms supply. True to style, in this cooperation Laffitte is definitely not a junior partner. In fact, if anything, he makes the bigger contribution by disrupting the network's communications, distracting its operatives, and devising a cunning plan to break into one of its hideouts, destroy its arms store, and free a hostage. During that operation, it is, again, Laffitte, the apolitical French (almost-) everyman who quite literally tells Winkelmann, the professional Stasi Kundschafter, what to do.

Moreover, Laffitte is not only clearly the single most important hero of "Insel des Todes." Occasionally, he is also intriguingly reminiscent of the single hero agent of *Visier I*. Thus, like Achim Detjen in the first version of *Visier*, Marcel Laffitte succeeds by subverting a secret organization from within. For Detjen, it was the old Nazi network; for Laffitte, it is the neofascist conspiracy ultimately run by the CIA. Like *Visier I*'s Detjen as well, Laffitte scores his biggest hit when he infiltrates a medieval castle that serves nefarious new purposes: in *Visier I*, a German castle is the site of a key meeting on rearming West Germany; for Laffitte in *Visier II*, its French equivalent provides camouflage for the neofascist organization's island lair. And both Lafitte and Detjen use old, hidden passageways to dupe their antagonists. Viewers with a good memory would have seen echoes in individual shots too: Lafitte exploring the French castle and a system of tunnels underneath it with the help of a flashlight visually recalls Detjen doing, in essence, the same at the West German site of the rearmament conference.

If Laffitte was notably central to the last *Visier* adventure, like no inhabitant of the West before, he may have been meant for greater things still. At the very end of "Insel des Todes," he learns from Winkelmann that his attempt to expose the whole conspiracy by providing a prosecutor with evidence of its crimes has failed. While he has killed most of the murderers of his family, the upper echelons and masterminds of the plot, including the head of the Italian secret service, are, in essence, getting away. He vows to "return to the Island of Death."

Yet *Visier* was discontinued.[20] But there may have been plans to go on and, in particular, to produce a sequel of this particular story. Laffitte's final vow, the fact that despite some setbacks the neofascist conspiracy was still essentially intact, and the curious anticlimactic detail that viewers see Laffitte mine its headquarters but not blow them up all point to such a possibility. Was there a plan to make the Laffitte character even more important, a sort of second Detjen, once again shifting the emphasis from a team of Kundschafter to a more or less single hero, as in *Visier I*? If so, it is fascinating that Laffitte was French, not German (whether East or West), bereft of any recognizable political attitude, and joined the struggle against the neofascists for motives of personal revenge. Was he meant to develop, under the tutelage of Stasi guidance, into a more conscious hero of socialist intelligence, pronouncedly internationalist since coming from a capitalist country in the West? Probably, we will never know.

What we do know is intriguing enough, however. If we look at *Visier I* and *II* as a whole, there were two clear changes in how the films depicted the East German Kundschafters' relationship with ordinary people in the West. In *Visier I*, the latter play a secondary role in that the Kundschafter hero has fairly little to do with them. While certainly never harming them, he also does not protect or rescue individuals. In the first two stories of *Visier II*, we see his successors from the Kundschafter team go out of their way to do just that. And, finally, in the last instalment of *Visier II,* and thus *Visier* in general, an ordinary man from the West turns into the central hero of the film, doing most of the real work of fighting the neofascist enemy, saving a member of the Kundschafter team, and, on the whole, clearly overshadowing the latter.

---

20. There was a subsequent spin-off, "Fire Dragon" (Feuerdrachen), and an unrealized plan for a second spin-off under the title "Jungle of Missiles" (Raketendschungel). See Haller, "Imaginations of Insecurity," 206.

Conclusion

Visier was a fantasy about noble East German agents heroically battling the Western Cold War Other on its own turf. It was unexceptional in its pronounced lack of realism, which is no strength of escapist film entertainment anywhere. Its politics were, generally speaking, unsurprising as well: the Cold War East was good, the West was bad; capitalism was corrupt and cynical, socialism Soviet-style (roughly speaking) was clean and humane; and finally, the legacies of fascism were alive in the West (especially West Germany), and not in the East (not even in East Germany). So far, so predictable.

Yet, as shown above, a closer, careful look reveals that *Visier*—even its often shoddily scripted second, team-based iteration—was a rich artifact of Cold War popular culture, with complex and even unexpected messages. Regarding complexity, the image of the heroic East German agent included a running comment of compensatory wish fulfillment. Here were ideal East German citizens doing their duty, deserving admiration at home, and yet also getting a fair slice of the capitalist good life abroad, out of reach for most of their compatriots. On top of that, these agents consistently punched above their weight vicariously for East Germany as a whole. Like Britain's James Bond, they served an at-best middling power and yet were doing major things in the world at large. And finally, perhaps most satisfyingly of all, they turned into gentle, benevolent guardian angels of hapless West German cousins, neatly reversing West Germany's claims of superiority.

But what emerges as *Visier*'s most unexpected turn also marked the end of the franchise: the elevation of an unpolitical, if heroic, Western everyman (from France, not West Germany) into the films' real key hero. Since *Visier* was not continued, we will never know whether this was meant to remain an exceptional move or to initiate a whole new framing of the franchise.

Acknowledgments

An early version of this article was presented at the "Narrating Cold Wars" Conference in November 2021, organized by the School of Communication and Film at Hong Kong Baptist University. I thank the organizers, commenters, and participants for feedback, encouragement, and an inspiring meeting.

To Whom Have We Been Talking? Naeem Mohaiemen's Fabulation of a People-to-Come

NOIT BANAI

Abstract

This essay considers Naeem Mohaiemen's three-channel video installation *Two Meetings and a Funeral* (2017) and performance-lecture *The Shortest Speech* (2019/2021) in order to expand the frameworks through which the Cold War might be understood from our contemporary perspective. It analyzes the technologies, techniques, and conventions of social assembly and public address embedded in and animating Mohaiemen's works and makes visible the problematics of imagining a "people" during the Non-Aligned Movement's historical context and in our neoliberal global capitalist present.

Keywords: Cold War, Non-Aligned Movement, Storytelling, Time-based media, Contemporary art

This essay analyzes Naeem Mohaiemen's three-channel film installation *Two Meetings and a Funeral* (2017) and his performance-lecture *The Shortest Speech* (2019/2021). These two mutually constitutive artistic interventions have contributed to destabilizing the dominant East-West Cold War binary by focusing on the role of the Non-Aligned Movement (NAM) as a third force—itself negotiating multiple, often conflicting, internal agendas—and, specifically, by considering Bangladesh's pivot from socialism to Islamism between the NAM Summit of 1973 in Algeria and the Organization of Islamic Cooperation (OIC) meeting of 1974 in Pakistan. The

https://doi.org/10.3998/gs.2437

115

critical conversation around *Two Meetings and a Funeral* and *The Shortest Speech* has predominantly focused on the unfinished, transversal histories and yet-to-be discovered archives of various liberation movements nested within the NAM and its ultimate failure to govern.[1] To develop these lines of thought in new directions, this essay brings to the fore the issues of communication and communicability, which were fundamental to the creation of an alternative socius and clearly preoccupied its leaders. By analyzing the technologies, techniques, and conventions of social assembly and public address embedded in and animating Mohaiemen's works, I would like to make visible the problematics of imagining a "people" during the NAM's historical context and in our neoliberal global capitalist present.

Two Meetings and a Funeral presciently begins with a temporal return to the day before as Sinnathamby Rajaratnam, the foreign minister of Singapore, takes the stage to address the delegates at the fourth NAM Summit on the fifth day (September 5–9, 1973, in Algiers):

> Yesterday, Mr. Chairman, for some reason, we had a technical breakdown. All the equipment that we are using to threaten the big powers is provided by them. They broke down and we could not communicate. We are all sitting here in a place made and built by the great powers. With that we cannot hold this conference. We sent telegrams to our home countries. We

I am grateful to Naeem Mohaiemen for his intellectual and artistic generosity.

1. Stephanie Bailey, "The Spectre(s) of *Non-Alignment(s)*," *di'van: A Journal of Accounts* 9 (March 9, 2021): 124–37; Noit Banai, "Documenta: Border as Form," *Artforum International* 56, no. 1 (September 2017): 302–5; Kaelen-Wilson Goldie, "Shifting Ground: On Stories and Archives in the Work of Naeem Mohaiemen," *Afterall* (Spring–Summer 2019): 67–77; Tom McDonough, "Incorrect History," *Texte zur Kunst*, no. 107 (September 2017): 163–65; Chris Moffat, "Dust and Debris in the Films of Naeem Mohaiemen," *Berfrois* (November 22, 2018); Vijay Prashad, "Naeem Mohaiemen's Tragic History of the 1970s Left," *Afterall* (Spring–Summer 2019): 59–67; Bilal Qureshi, "Naeem Mohaiemen's Cinematic Resistance," *Film Quarterly* 71, no. 2 (Winter 2017): 61–64; Eli Rudavsky, "Two Meetings and a Funeral," *Chicago Review*, February 4, 2020.

had to send one to Singapore; it had to go to Paris, London, Singapore. If they turn it off, we are lost.[2]

Speaking to the mostly empty hall, sparsely populated with distracted participants—many of whom have taken their headphones off—Rajaratnam inveighs against the infrastructural dependance on the superpowers while punctuating the air with hand gestures that suggest the many directions in which information had to travel to reach its intended audience (fig. 1).

At the crux of his reflections is what Stephanie Bailey has justly characterized as NAM's problematic "reliance on the developed world for technological expertise, knowledge and equipment" or a "global economic system to which every nation-state in the movement were inevitably connected and often indebted."[3] Yet these deliberations on vexed entanglements and their uneven distribution of power are only a prologue to Rajaratnam's coup de

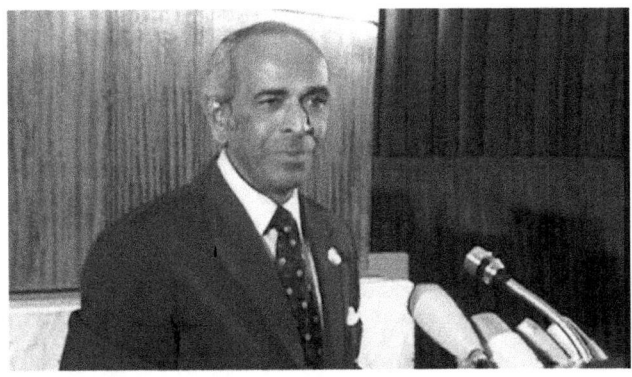

Figure 1: Sinnathamby Rajaratnam speaking at the fourth NAM Summit, Algiers, September 9, 1973. Naeem Mohaiemen, *Two Meetings and a Funeral*, 2017. Courtesy of artist.

2. Naeem Mohaiemen, dir., *Two Meetings and a Funeral* (Kassel, Germany: Documenta Und Museum Fridericianum gGmbH), 2017, 1:03–2:03.
3. Bailey, "The Spectre(s) of *Non-Alignment(s)*," 128.

grâce: "Mr. President, this is almost the tail end of nearly one week of speeches and deliberations. . . . I've asked myself: *To whom have we been communicating this last one week? To whom have we been talking? To ourselves? Or, to the 2 billion people we are supposed to represent?* What is it that they would require of our Non-Aligned Conference? Have we provided them the solutions that our peoples have been asking?"[4] While underlining the need for concrete solutions for NAM's diverse constituencies, Rajaratnam also expresses underlying anxiety about the communicability of images, ideas, and experiences through which to constitute an image of a "people" that would be inclusive of all the communities under the NAM aegis but distinct from the dominant ones offered by the capitalist and communist superpowers.

This issue had been crystallizing since the Bandung Conference of 1955 and the formation of the NAM in Belgrade in 1961. What distinguishes the 1973 conference from those earlier moments, as Mohaiemen asserts in *The Shortest Speech*, is that many of the liberation movements had come to power and were governing newly decolonized nation-states. Their dilemma was couched in the double challenge of affirming and strengthening the sovereignty of singular nation-states—many of whom had been colonized by Western powers and contained varied ethnic, religious, and linguistic groups—while also developing a collective ideology meant to supersede the modern power structures that had oppressed so many in the name of nation, class, and race. The political theorist Étienne Balibar, writing about the tensions within the two rival blocs during the Cold War, observes that "each of these presented itself as supranational, as an internationalism, for there was a liberal internationalism as there was a socialist internationalism. It is, however, doubtful whether the 'blocs,' inasmuch as they were mutually exclusive and organized around state constructions, found any other cement for their internationalism than an expanded, loosed-up form of nationalism."[5]

---

4. Mohaiemen, 2017, 5:23–7:10. Emphasis is mine.
5. Étienne Balibar, "Ambiguous Identities," in *Politics and the Other Scene* (London and New York: Verso, 2002), 59.

The same operative paradox is found in the NAM, a medley of diverse nation-states espousing different ideologies, trying to extend themselves beyond the structures and power relations that had been the basis for colonial empires toward a supranational framework. Because of the colonial legacy, these nation-states and their peoples had long been rendered the "exteriority" to Europe's "internal" population and their "particularity" was not—and is still arguably not—integrated into Western "universality." One of the underlying questions raised by Mohaiemen is whether this structural model and its use of race, class, ethnicity, religion, and gender to fundamentally differentiate between "us" and "them" would become the prototype for NAM or if an alternative vision of identity and belonging could be created via third worldism. As if to underline this dilemma and the complicated relay of identifications (individual, nation, supranational federation of NAM), Mohaiemen repeatedly transports the viewer to the Palace of Nations in Chéraga, just outside of Algiers, where delegates are installed behind plaques bearing the name of their country in English and Arabic, some with newly minted flags perched on their tables, listening (or ignoring) charismatic individuals such as Fidel Castro, Yasser Arafat, Anwar Sadat, Indira Gandhi making passionate appeals to transnational solidarity (figs. 2 and 3).

The footage gathered by Mohaiemen in the archives of Algerian National TV conspicuously shows the army of cameramen, photographers, translators, and typists hired to capture and circulate these leaders' words and images.

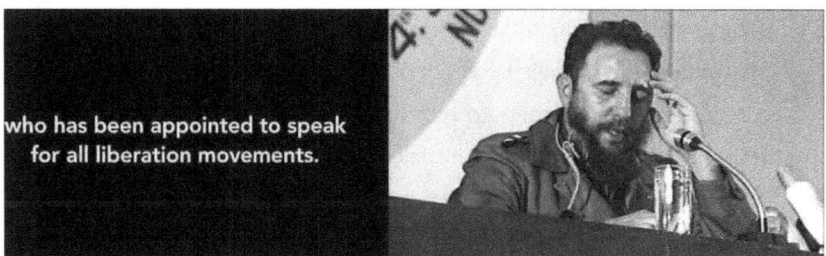

who has been appointed to speak for all liberation movements.

Figure 2: Fidel Castro speaking at the fourth NAM Summit, Algiers, September 8, 1973. Naeem Mohaiemen, *Two Meetings and a Funeral*, 2017. Courtesy of artist.

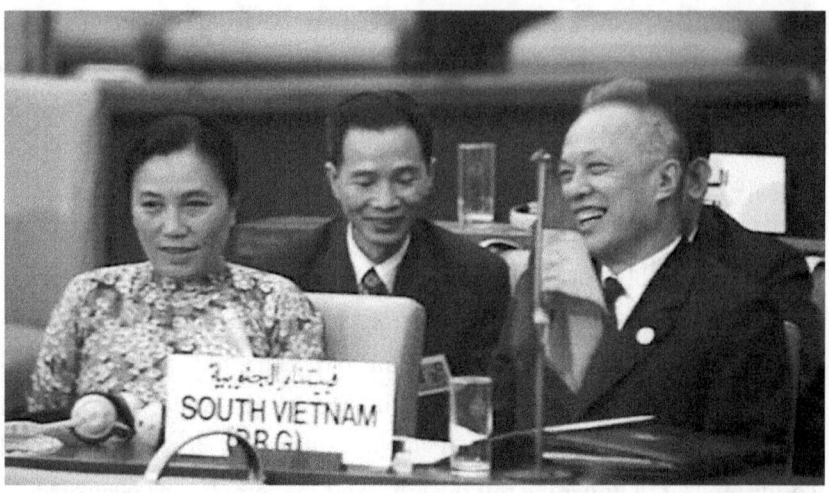

Figure 3: The delegation from South Vietnam (PRG) at the fourth NAM Summit, Algiers, September 8, 1973. Naeem Mohaiemen, *Two Meetings and a Funeral*, 2017. Courtesy of artist.

They mill around in the conference hall searching for photogenic moments— offered willingly by the likes of Castro—and have installed their equipment at the podium, ready for sound bites and sweeping shots of the venue (fig. 4).

The conference participants are intensely aware of this ubiquitous media presence and performatively shake hands and exchange embraces to satisfy the lenses and scribes. They also frequently turn a blind eye to the media when it is at a distance and assisted by newer TV camera technology, preferring to make small talk among themselves or stare blankly into space after many days of epic speeches. It is not only the historicity of the event that is documented but also the fundamental issue succinctly articulated by Rajaratnam when he asks, "To whom have we been talking?" In these montaged episodes, I suggest that we are not simply witnessing addresses made to a ready-made or already determined "people"; rather, in the pressing context of self-governance and a planetary communication system, Mohaiemen makes it retrospectively evident that there was an exigency to simultaneously keep imagining *a people-to-come*. On what

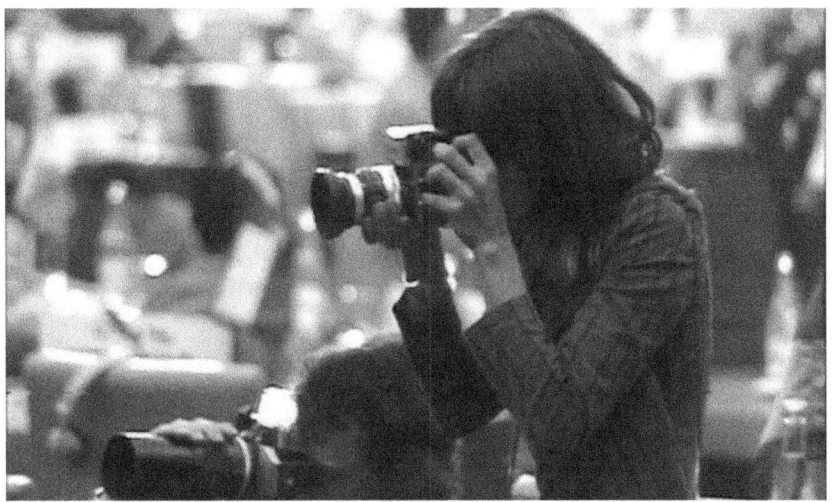

Figure 4: Camerawoman at the fourth NAM Summit, Algiers, September 5–9, 1973. Naeem Mohaiemen, *Two Meetings and a Funeral*, 2017. Courtesy of artist.

expressive grounds would this NAM collective be envisaged? What would be the relationship between discursive, visual, and political representation? And what would be the most generative means to broadly disseminate an image of this "people" in a political space enmeshed with an intensified media ecology?

Now empty, this cavernous hall appears as a relic of a political project that can only be activated in the present by contemporary narrators who contribute discursive fragments to this history's eventual, if not predetermined, failure. Vijay Prashad, Samia Zennadi, Attef Berredjem, Amirul Islam, and Zonayed Saki take us on a visual narrative journey through Algiers, New York, and Dhaka in search of traces of this not-so-distant past (fig. 5).

Theirs is a project of personal and collective memory, which simultaneously helps draw the parameters of our current inability to conjure an alternative political imaginary. Anecdotes about Yasser Arafat's preferred hotel, reflections on the United Nation's defunct index card filing system, and a visit to the Bangabandhu International Conference Center make evident

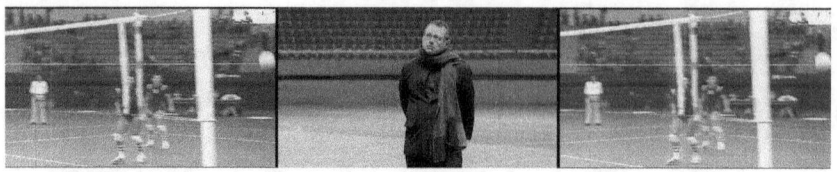

Figure 5: Vijay Prashad at the stadium of La Coupole d'Alger juxtaposed with earlier sports competitions, Dely Ibrahim, Algeria. Naeem Mohaiemen, *Two Meetings and a Funeral*, 2017. Courtesy of artist.

the international scope of the sites, apparatuses, and institutions through which the identity of the NAM community was negotiated. As spectators of a cinematic experience woven from dialectical images and a hypnotic electronic soundtrack, we only learn who these protagonists are at the end of the film, when the credits roll, leaving us in the hands of a cast of unknown, unreliable storytellers who reflect on the NAM's emancipatory ambitions, the reasons for its failure, and the complex situations left in its wake in Algeria and Bangladesh. Juxtaposed with the stirring rhetorical performances by the NAM dignitaries in 1973, the improvised exchanges of our contemporary narrators are a stark counterpoint: historic speeches in the name of a dreamed future, on the one hand, and fragmented streams of consciousness about a miscarried past on the other. It is as if, Mohaiemen suggests, there has been a critical rupture not only in terms of a political imaginary but also in forms of enunciation and public address.

To stress this point, Mohaiemen takes us to the publicity circuit—the television appearances at which NAM leaders addressed the masses via one of the most advanced forms of technology at the time. We observe Marcelino dos Santos, the Mozambican poet and revolutionary who served as the president of the Frente de Libertação de Moçambique (Front for the Liberation of Mozambique, FRELIMO) from 1969 to 1977, explain the significance of FRELIMO'S participation at the NAM summit in Algiers (fig. 6).

Speaking in French, he lucidly articulates the stakes surrounding NAM's control of their natural resources and the connection between economic

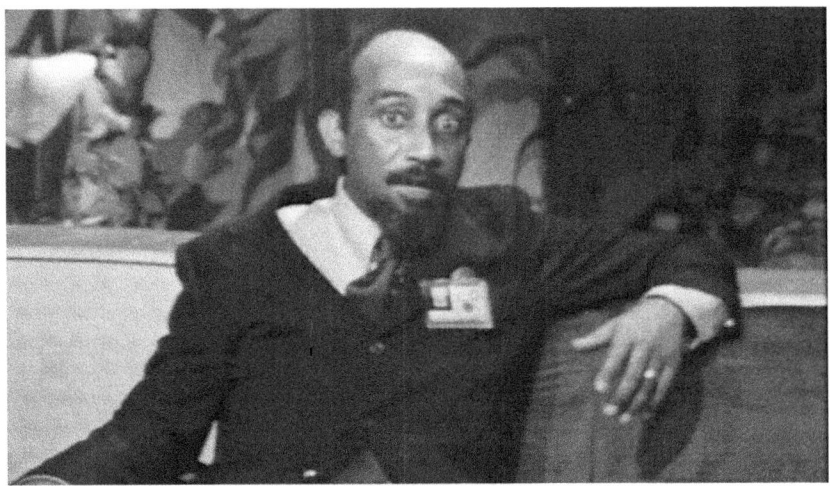

Figure 6: Marcelino dos Santos giving a televised interview about Frente de Libertação de Moçambique's participation at the fourth NAM Summit. Naeem Mohaiemen, *Two Meetings and a Funeral*, 2017. Courtesy of artist.

and political independence. The argument is not diluted either in content or form. What is evident is that dos Santos considers television a prime tool to disseminate the NAM's demands for equality and self-determination. For the revolutionary, rational linguistic debate goes hand in hand with a televisual image: transmitting a message of self-governance is not at odds with what art historian David Joselit has termed "television's privatization of public speech."[6] On the contrary, because of television's increasing collusion with dominant (Western capitalist) interests, it is the perfect vehicle through which to launch an appeal for an alternative public. Yet, there is something else animating dos Santos's overture: a realm of living speech, which I argue was still extant, if already eroded, by 1973. This is a leader who is speaking to the masses via television but has not altered his conceptualization and register of address to the structural parameters of the medium. Or in an adaptation of Marshall Mcluhan's famous dictum, "the

---

6. David Joselit, *Feedback: Television Against Democracy* (Cambridge, MA: MIT Press, 2007), xi.

medium *does not* become the message."[7] As he is soliciting the judgment of a public that had been colonized by Western powers (in the case of Mozambique by the Portuguese), and thus effectively stripped of its right to question, he is also imagining a people-to-come for whom judgment would be an intrinsic human right.

Mohaiemen further explores the potentially radical role of television in the context of Bangladesh's reversal of affiliation from socialism to Islamism. The artist accompanies the Bangladeshi politician Zonayed Saki to the Bangabandhu International Conference Center, a venue built for an NAM conference ultimately held elsewhere that is now available for corporate rental.[8] We hear Saki explain to a well-wisher that he is there to shoot a documentary while attendees of a trade fair circulate in the background. Upon entering the building, the camera is suddenly turned off, and an exchange ensues between Saki, Mohaiemen, and the venue's chief of security. Captured by a live microphone, the artist is heard trying to clarify that he is making a documentary film about the NAM while the official insists that no television is allowed and that NAM is dead. In this misunderstanding over media forms—ironically linked to the NAM's existence—we grasp that neoliberal global capitalism renders certain televised speech as suspect. Allowing Saki, recognized immediately by the official as a left-leaning politician, to

---

7. In his analysis of the media through which information and knowledge is communicated to the mass public, Marshall McLuhan famously stated that "the medium is the message" in *Understanding Media: The Extensions of Man* (Cambridge, MA: MIT Press, 1994).

8. Bangabandhu International Conference Center (BICC) was renamed in 2013 to replace its former title, Bangladesh China Friendship Conference Center. Funded by the People's Republic of China and designed by the Beijing Institute of Architectural Designs and Research, it was completed in 2001 with the intention of hosting an NAM summit that was ultimately held in Malaysia in 2003. Under Bangladesh's newly elected government of prime minister Khaleda Zia, the finance minister Saifur Rahman "had termed NAM a dead horse and insisted that Bangladesh cannot afford to spend millions of dollars for its burial." See 'Non-Aligned Summit in Dhaka Next Year Put Off: Official," Zee News, October 16, 2001, https://zeenews.india.com/news/south-asia/ nonaligned-summit-in-dhaka-next-year-put-off-official_23287.html.

occupy an enunciative space that might destabilize the consumerist image of the venue or represent a different social relation to capital is inconceivable and the camera is summarily turned off.

It is telling that Mohaiemen's own voice is heard twice in *Two Meetings and a Funeral*, both times as an off-screen presence, when he negotiates camera access at the Bangabandhu International Conference Center and while pitching questions to Vijay Prashad at the stadium of La Coupole d'Alger. In contrast, the artist is a central protagonist in the performance-lecture "The Shortest Speech," his soft, accented narration intersecting with the montage of images and weaving another layer of interpretation around *Two Meetings and a Funeral*. Framing Rajaratnam's speech as a minoritarian moment of doubt that punctuates an otherwise affirmative conference, Mohaiemen moves between the visible apathy and inattentiveness of the remaining participants and the momentous events that await them. The Afghan delegate, Mohaiemen remarks, is lazily smoking a cigarette, unaware that his socialist government will be overthrown by the Soviet Union in six years. Anwar Sadat, pipe in mouth, cannot foresee that breaking with the Arab Pact and normalizing relations with Israel will lead to his assassination in eight years' time. Salvador Allende, visibly absent in Algiers, will be assassinated only two days after the end of the conference, on September 11, 1973—what the film calls "the Other 9/11."

Mohaiemen embeds Rajaratnam's speech within these dramatic elisions and ruptures between a present (in the past) and a future (in the past, present, and a time to come). It is only this short-story writer turned politician who gives the shortest speech, who can foresee the new problem emerging in the wake of the signing of the SALT I treaty between the United States and the Soviet Union in 1972. In Mohaiemen's telling, this is a "pivot moment in third world solidarity." . . . "If non-alignment was partially premised on socialist solidarity outside of alignment with either of the Cold War powers. . . . What happens when your entire reason for existence is partially taken away by a treaty between two superpowers, what remains after opposition? . . . We all agree on what we are against, but what exactly are we for?"[9] Here, indeed, is Mohaiemen—a historical subject born into these geopolitical shifts—speaking *through* the open questions elicited by Rajaratnam's speech. Stated differently, the message of the short-story writer (Rajaratnam) is delivered by a contemporary storyteller (Mohaiemen) who personally experienced the outcomes of the NAM's failure to conjure an image of a people that would be in sync with both a mandate of governance and revolution.

Importantly, in confronting this historical past via subjective memory, the artist assumes the right to fiction or what Gilles Deleuze has called, after Henri Bergson, "fabulation."[10] According to Deleuze, it is fabulation or "legending" that allows "a minority discourse, with one or many speakers, to take shape."[11] In Deleuze's parlance, legending is an act of counternarration that renders visible the fictions that have become accepted as orthodoxy *without* presuming to offer another concretized or completed ideal. It is open-ended, processual, and requires the participation of a collectivity to vitalize and acknowledge it. Importantly, "to catch someone in the act of legending is to catch the movement of constitution of a people. A people isn't something already there. A people, in a way, is what's missing."[12] Thus, to be made aware of the legending that is occurring is to be part of a process of generating new truths that are "not already out there" but "have to be created in every domain."[13]

Mohaiemen's fabulating, legending, or storytelling—in spoken form and time-based media—takes a slightly different register from the written format of the short story. While the artist "performs" his "lecture" from an already drafted textual piece and an attendant visual presentation of still and

9. Naeem Mohaiemen, *The Shortest Speech,* performance-lecture delivered at the Narrating Cold Wars Conference, Hong Kong Baptist University, November 11, 2021, 28:25–29:04.
10. Gilles Deleuze, *Negotiations,* trans. Martin Joughin (New York: Columbia University Press), 125.
11. Deleuze, *Negotiations,* 125.
12. Deleuze, 126. Emphasis is mine.
13. Deleuze, 126.

moving images, it is the oral delivery that is crucial. According to Walter Benjamin, storytelling, at its inception, was linked to a collective public and the shared experiences that bind them together. For Benjamin, "the storyteller takes what he tells from experience—his own or that reported by others. And he in turn makes it the experience of those who are listening to his tale."[14] In assuming this role, Mohaiemen generates a reciprocal relation with a processual public: we *become* listeners while a shared history is emerging during the live process of performance. Here, the scene at the Bangabandhu International Conference Center is decisive. As the camera is switched off, the head of security aggressively quizzes Mohaiemen: "How many years are you in media? Where do you live in Dhaka? Where did you study? All your schooling was in America? What year at Dhaka college? . . . I also went to Dhaka college. Are you sure you went there? I remember everything."[15] What surfaces in this back and forth is the way in which the head of security tries to assess Mohaiemen's status through his biography and, specifically, his schooling: local or foreign? public or private? In their exchange, class emerges as the central criterion for interpersonal differentiation, and the power of memory is claimed as an arbiter of historical truth. As the scene unfolds, one of Mohaiemen's concerns—to understand the internal disintegration of the NAM and trace how and why Bangladesh's initial embrace of state socialism was renounced—comes to the fore. Even as the NAM nations committed themselves to decolonization and to countering the Western and Eastern blocs' model of a "people," they were not able to eradicate certain structural elements, class among them. In *Essays Critical and Clinical*, Deleuze's notion of the "people to come" is expanded: "A minor people, eternally minor, taken up in a becoming-revolutionary. Perhaps it exists only in the atoms of the writer, a bastard people, inferior,

---

14. Walter Benjamin, "The Storyteller: Reflections on the Works of Nikolai Leskov," in *The Novel: An Anthology of Criticism and Theory 1900–2000*, ed. Dorothy J. Hale (Malden, MA: Blackwell Publishing, 2006), 364.
15. Mohaiemen, *The Shortest Speech*, 1:14:02–1:14:45.

dominated, always in becoming, always incomplete. Bastard no longer designates a familial state, but the process of drift of the races."[16]

While power differentials may persist within a society, to remain revolutionary, concepts such as race, class, ethnicity, religion, and gender must remain in permanent drift, without ever fully concretizing. This, in turn, leads us to ask, with Mohaiemen: Was a *doxa*—or majority prejudice—established when the NAM nations started to govern? Did they compromise their revolutionary goals—and the image of the "people to come"—when they assumed genuine political power? Crucially, is it possible to simultaneously constitute both a people and people to come?

As a way to maintain a constitutive openness—or drift—in the present, Mohaiemen regularly makes the audience aware of his narrative's "constructedness" and its potential for multiplicity. For instance, commenting on the general distraction in the conference hall, he states, "You wonder when the break will come, for coffee and the endless cigarette; *but an anachronism is revealed in the way I narrate a script of impatience. Coffee and cigarettes are what I insert as the elements that are awaiting impatient delegates.* But this is 1973, a time when you can freely smoke inside airplanes, in lobbies, in offices, cafeterias, and certainly at a conference."[17]

By exposing this speculative condition and disrupting the possibility of a seamless history, Mohaiemen clarifies that he can only engage in this act of mnemonic reconstruction from a contingent and ever-becoming present. Paradoxically, it is precisely this narrative partiality—or what Donna Haraway has called his "situated knowledge"—that lays the groundwork for a collective experience in today's global context. Imagining a people means developing "the ability partially to translate knowledges among very different and power differentiated communities."[18] For Mohaiemen, such a trans-

---

16. Gilles Deleuze, *Essays Critical and Clinical*, trans. Daniel W. Smith and Michael A. Greco (Minneapolis: University of Minnesota Press, 1997), 4.

17. Mohaiemen, *The Shortest Speech*, 15:12–15:42. Emphasis is mine.

18. Donna Haraway, "Situated Knowledges: The Science Question in Feminism and the Privilege of Partial Perspective," *Feminist Studies* 14, no. 3 (Autumn, 1988): 580.

lation among "power differentiated communities" goes hand in hand with an intentional evasion of authority: he is not the omniscient Bard who acts as the definitive repository for a completed history but, on the contrary, he is the fabulator who carves out a temporary, performative locus that any of us could also potentially occupy, embody, and animate.

In the dynamic of speeches, stories and spaces of assembly, address and publicity that constitute *Two Meetings and a Funeral* and *The Shortest Speech*, Mohaiemen poses the question: What modes of discourse and means of communication were necessary to fabricating the NAM's emancipatory image of the "people" circa 1973–1974, and, by extension, what are the grounds for such an imaginary today? The complex visual, sonic, and aural *fabulations, legending, storytelling* through which these artworks materialize suggest that politics and art operate on a shared terrain in which knowledge and power are permanently negotiated, dissolved, reformulated, and translated. Animating both is a kernel of fiction, a polyphony of voices, and sediments of situated knowledges that are in search of a "people-to-come."

The Man without a Country

British Imperial Nostalgia in *Ferry to Hong Kong* (1959)

KENNY K. K. NG

Abstract

On New Year's Eve 1959, *Ferry to Hong Kong* was screened at the Lee Theatre and the Astor in Hong Kong. Produced by Rank as its first CinemaScope feature, the big-budget movie tells the real-life tale of Steven Ragan (he was also known as Michael Patrick O'Brien), a stateless drifter who was stuck for ten months on the ferry sailing between Hong Kong and Macau from September 18, 1952, to July 30, 1953. The British film was Rank's major Anglo-American joint venture of the year. Positioned within Cold War contexts, *Ferry to Hong Kong* could be seen as a British cultural-diplomatic response through cinematic soft power to reestablish national assurance on Asian Cold War fronts, following the 1956 Suez Canal debacle that witnessed the death of Britain's imperial might at the hands of the Eisenhower administration. Unlike such vaunted Hollywood pictures as *Soldier of Fortune* (1955), *Love Is a Many-Splendored Thing* (1955), and *The World of Suzie Wong* (1960), which imagined the incursions of American white knights into Hong Kong (as a stand-in for China), *Ferry to Hong Kong* conveyed imperial nostalgia and loss. The film turns the antihero into a paragon of British gallantry who saves the passengers and refugees from the hands of Chinese (Communist) pirates. The sinking ferryboat is the traumatic device used to recall British naval war stories and retell romantic and narcissistic tales of British valor and international influence. More than an adventure of a vagabond, *Ferry to Hong Kong* was an espionage thriller in uneasy disguise. The film preceded

Gilbert's three James Bond films, all of which affirmed the power of the individual in cracking transboundary networks of espionage and political intrigue.

Keywords: Anglo-American coproduction, Cold War tourism, Hong Kong cinema, Orientalism, Orson Welles

On New Year's Eve 1959, *Ferry to Hong Kong* was screened at the Lee Theatre and the Astor in Hong Kong. Produced by Rank Organization as its first CinemaScope feature, and directed by Lewis Gilbert, the British film was shot on location in Hong Kong and Macau from October 1958 to February 1959.[1] The big-budget film tells the real-life tale of Steven Ragan (also known as Michael Patrick O'Brien), a stateless drifter who was stuck for ten months (315 days) between 1952 and 1953 on the ferry shuttling between Hong Kong and Macau. Born an Austro-Hungarian, Ragan had previously worked in a Shanghai bar before he escaped from China, after the Communist takeover, and set foot in Macau in 1952. When his permit was revoked in the former Portuguese colony, he sneaked onto a Hong Kong-Macau ferry to avoid deportation to mainland China.[2] The international press keenly reported on his unusual but hard-to-be-vindicated experience, which was subsequently fictionalized by Simon Kent (the penname of Max Catto) in the book *Ferry to Hongkong*.[3]

---

1. A. H. Weiler, "Brand-New 'Scent' on the Todd Roster–Kelly's 'Gentleman'–Addenda," *New York Times*, September 28, 1958; "Orson Welles Here to Make Picture: Will Play Part of Skipper in *Ferry to Hongkong*: Staying Four Months," *South China Morning Post*, November 17, 1958, 6.
2. Henry R. Lieberman, "Ferry Voyager Leaves Hong Kong After 23, 680-Mile Trip to Nowhere: O'Brien's 315-Day Yo-Yoing Is Ended as Police See Him Off on Plane," *New York Times*, July 31, 1953; and Daniel Hånberg Alonso, "The Man without a Country," Medium, July 8, 2020, https://medium.com/@danielhanberg/the-man-without-a-country-df0044dddd09.
3. Simon Kent, *Ferry to Hongkong* (London: Hutchinson, 1958).

In the film adaptation, the Hong Kong police deport Mark Conrad (played by Curt Jurgens), a drunken Anglo-Austrian exile, tossing him aboard a ferry with a one-way ticket to Macau. When Macau rejects him as an undesirable, Conrad is condemned to be a man without a country. Taking the ferry as his makeshift home, he finds himself in a romantic encounter with Liz Ferrers (Sylvia Syms), a kindhearted schoolteacher cruising daily with her Chinese students on the passenger boat.

Ferry to Hong Kong was marketed as an "international film" targeting the US market.[4] Rank contracted British box-office star Curt Jurgens and famed American director and actor Orson Welles (fig. 1).

But British critics unanimously disapproved of Orson Welles, who played Captain Hart of the ferryboat, as giving his worst-ever performance.[5] They ridiculed Welles's deliberately bogus upper-class accent as a caricatured combination of Winston Churchill and the American comedian Jackie Gleason.[6] Welles intended to "bluster to the extent that he [got] many unintentional laughs,"[7] but his outrageous performance simply made audiences gasp in disbelief. The collaborations between Gilbert, Welles, and Jurgens did not work out smoothly. Gilbert described the shooting experience as outright "nightmarish," mainly because of Welles's arrogance.[8] Jurgens and Welles swapped their roles before filming: Welles played the captain of the ferryboat and Jurgens the tramp.[9] Welles had an interest to make the film a comedy, and often changed his own lines in filming, while Jurgens played

---

4. Sue Harper and Vincent Porter, *British Cinema in the 1950s: The Decline of Deference* (Oxford: Oxford University Press, 2003), 52–55.

5. "Orson Welles in a Film Typhoon: *Ferry to Hong Kong*," *Times*, July 1, 1959; "London Critics Flay Orson Welles," *China Mail*, July 3, 1959.

6. John Howard Reid, *150 Finest Films of the Fifties* (Morrisville: Lulu.com, 2015), 123.

7. "Flashes Review: *Ferry to Hong Kong*," Boxoffice, May 8, 1961, https://lantern.mediahist.org/catalog/boxofficeaprjun179boxo_0252.

8. Kirsty Young, interview with Lewis Gilbert, prod. Leanne Buckle, BBC Radio 4, June 25, 2010, https://www.bbc.co.uk/programmes/b00sqkg8.

9. Lewis Gilbert, *All My Flashbacks: The Autobiography of Lewis Gilbert* (London: Reynolds & Hearn, 2010), 187–96.

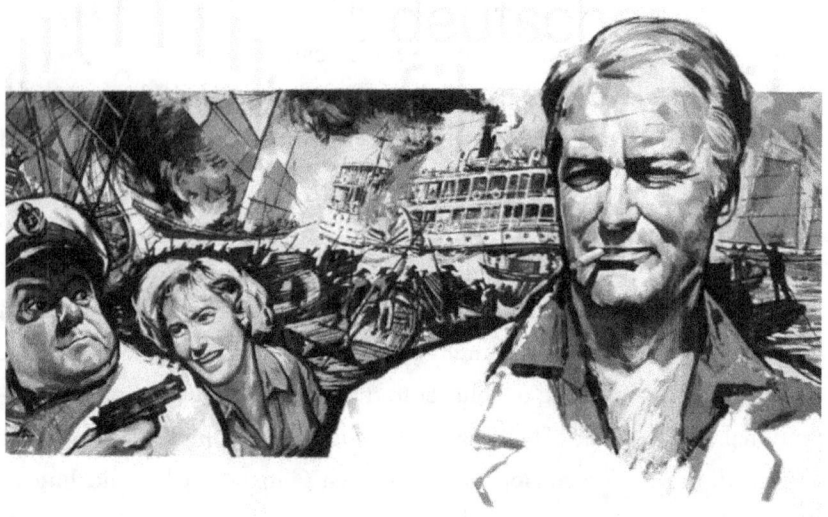

GALA WORLD PREMIERE

ODEON THEATRE LEICESTER SQUARE

July 2nd 1959 at 8.30 p.m.

in aid of The Newspaper Press Fund

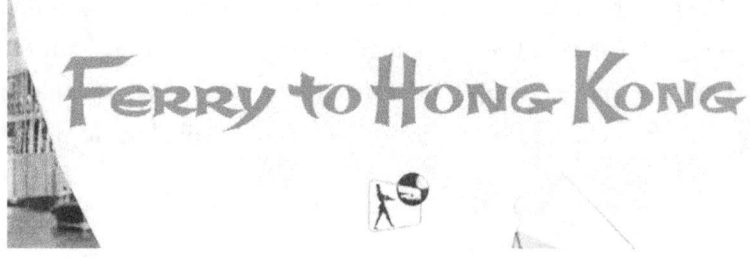

FERRY to HONG KONG

Figure 1: Program of the gala world premiere of *Ferry to Hong Kong* in London on July 2, 1959.
Source: "*Ferry to Hong Kong* (1959) Programm der Gala World Premiere, London (GB)," Curt Jürgens: The Bequest, Deutsches Filminstitut & Film-museum, 1959, https://curdjuergens.deutsches-filminstitut.de/nachlass/ferry-to-hong-kong-1959-programm-der-gala-world-premiere-london-gb/. Reproduced by permission of Deutsches Filminstitut & Filmmuseum.

the role seriously.[10] Jurgens felt that his star status was being undermined by Welles and threatened to walk out.[11] The British and American stars hated each other (fig. 2).

In Gilbert's memoirs, Welles on the set insisted on wearing a false nose, at one point holding up shooting for two days because he lost the correct false nose (fig. 3).[12]

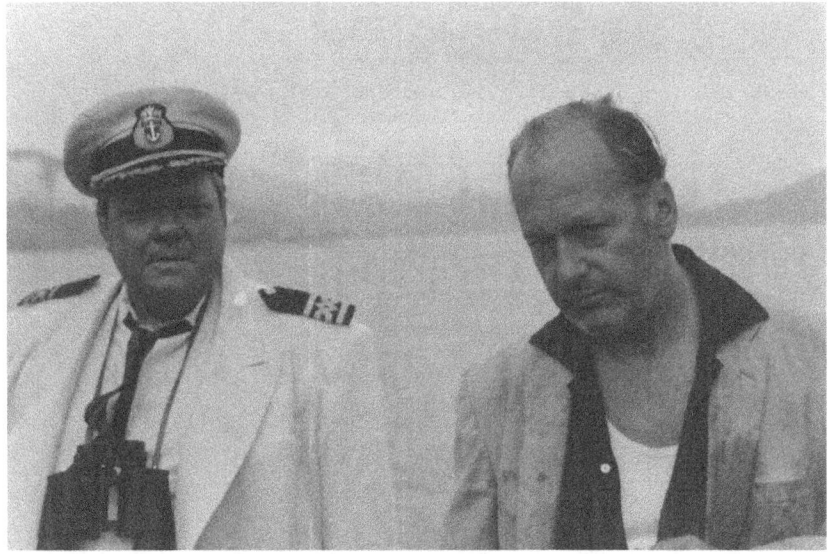

Figure 2: Orson Welles and Curt Jurgens on the set of *Ferry to Hong Kong.*
Source: "*Ferry to Hong Kong* (1959) *Werkfoto 4*," Curt Jürgens: The Bequest, Deutsches Filminstitut & Filmmuseum, 1959, https://curdjuergens. deutsches-filminstitut.de/nachlass/ferry-to-hong-kong-1959-werkfoto-4/. Reproduced by permission of Deutsches Filminstitut & Filmmuseum.

---

10. Chris Welles Feder, *In My Father's Shadow: A Daughter Remembers Orson Welles* (Chapel Hill, NC: Algonquin Books, 2011), 190–200; Orson Welles and Peter Bogdanovich, *This Is Orson Welles* (New York: HarperCollins, 1993), 264–66; Samuel Wilson, "A Wild World of Cinema: *Ferry to Hong Kong* (1959)," MONDO 70, December 15, 2017, https://mondo70.blogspot.com/2017/12/ferry-to-hong-kong-1959.html.
11. Quoted from Reid, *150 Finest Films of the Fifties*, 122.
12. Tony Sloman, interview with Lewis Gilbert, National Film Theatre Audiotape, October 23, 1995, quoted from Harper and Porter, *British Cinema in the 1950s*, 55.

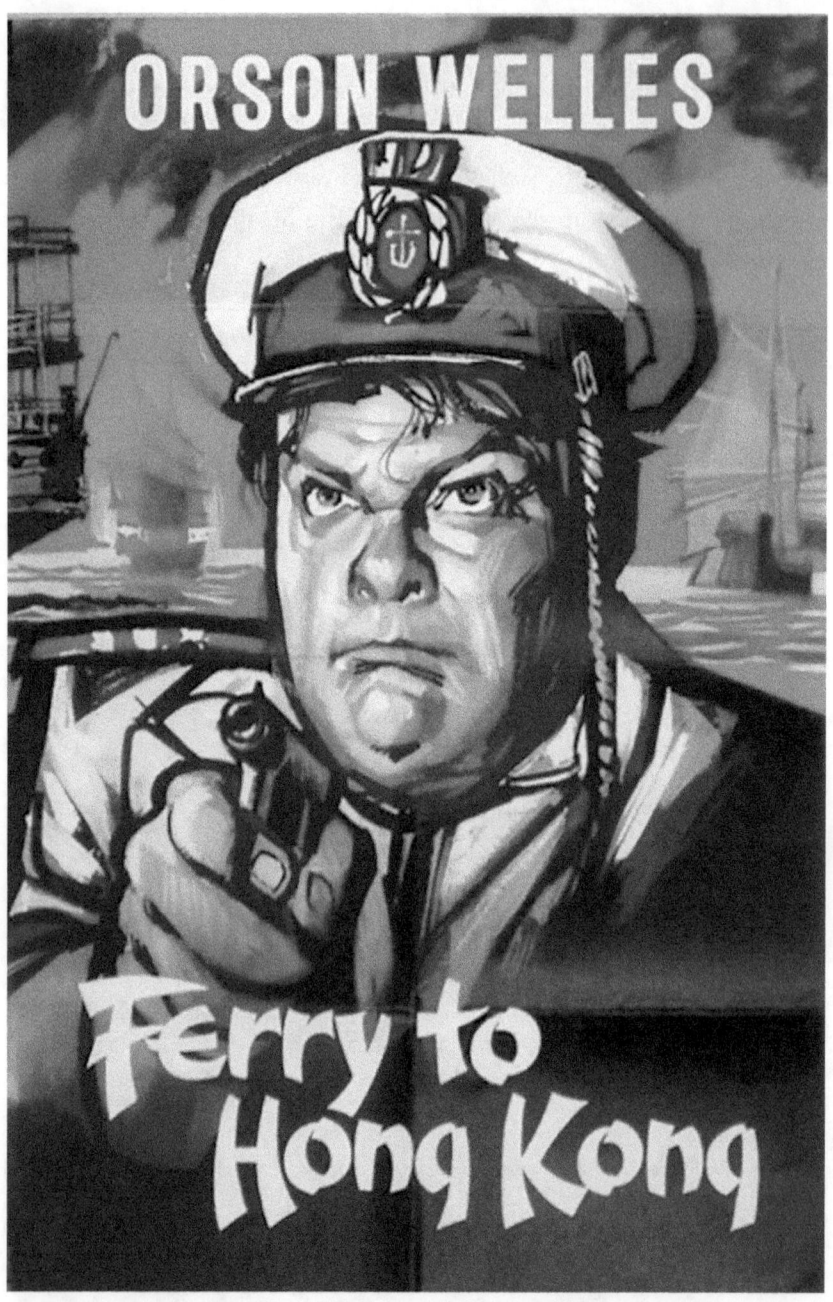

Figure 3: UK film poster of Orson Welles in *Ferry to Hong Kong*.
Source: Brian Bysouth, *Ferry to Hong Kong* (UK Release Poster), 1959,
Gregory J. Edwards on eBay, https://www.ebay.com/itm/120680537769.

He spoke with a bizarre voice with the false nose, and post-synching of his dialogues word by word only added to the weird effect.[13] Notably, the disservice Welles's overwhelming presence in the crew did to the filmmaking outweighed the value his fame was meant to add to the picture.[14]

In Hong Kong, David Lewin of *China Mail* conducted a four-installment interview series entitled "Third Man in Hong Kong" to echo *The Third Man* (dir. Carol Reed, 1949), an espionage noir set in postwar occupied Vienna. In the film, Welles portrayed the American racketeer Harry Lime, a fugitive sought after by the British military for trafficking in bogus penicillin to make a fortune.[15] Making allusions to the Cold War, Welles remarked on Hong Kong as the "new Third Man territory" and the "free, wide-open city."[16] Taken as he was by the perilous charm of the underground world in Britain's last colonial outpost, Welles viewed the city's illegal businesses as "all part of the Third Man style,"[17] notoriously as a haven for crime and adventure.

Gilbert's reminiscences present an unruly Welles, who could not work with his British partners. Forget about the charisma of Welles and his cinematic classic *Citizen Kane* (1941). I take the untold story of Welles in Hong Kong

---

13. Gilbert, *All My Flashbacks*, 189–91.
14. Welles later revealed in an interview after the filming that he thought *Ferry to Hong Kong* had proved to be "a disappointing B movie," one which "he wished to put out of his mind as soon as possible." Peter Moss, *No Babylon: A Hong Kong Scrapbook* (New York: iUniverse, 2006), 142.
15. Graham Greene wrote the screenplay for *The Third Man*, and his creation of the Harry Lime character was said to have been inspired by Kim Philby, the notorious British double agent for the Soviet Union. There is no question that Greene knew Philby, as they had worked in the same unit for British wartime intelligence. Tony Shaw believes that "Philby undoubtedly provided inspiration for the persona of Harry Lime." See Shaw, *British Cinema and the Cold War: The State, Propaganda and Consensus* (London: I. B. Tauris, 2001), 28. The unmasking of Philby as a Soviet "mole" inside MI6 provided the model for John le Carré's most famous espionage novel, *Tinker, Tailor, Soldier, Spy* (1974).
16. David Lewin, "Third Man in Hong Kong Part 1: Crime Is Inscrutable," *China Mail*, February 28, 1959. Welles was curious to know about the business of smuggling "human snakes" (gang slang for illegal immigrants) from China to Hong Kong.
17. "Harry Lime has gone up in the world," Welles told his interviewer. "He would still be making fast money: from smuggling illegal immigrants from Communist China . . . from dealing with dope . . . from trafficking in gold." Lewin, "Third Man in Hong Kong Part 1."

and, more importantly, the uncomfortable cooperation between the British and American filmmakers in making *Ferry to Hong Kong* as emblematic of a rift in the Anglo-American alliance to defend Hong Kong against Communist China's alarming aggression. The Anglo-American cinematic collaboration throws light on the uneasy relations between the two major Western Cold War allies, one a collapsed empire and the other an ascending global power.

On Asian Cold War fronts, the defense of Hong Kong had serious geopolitical and international implications for the United States. As Chi-kwan Mark contends, throughout the 1950s when conflicts in Korea, Indo-China, and the Taiwan Straits raised the specter of a Sino-American war in Asia, Hong Kong became a crucial concern in US foreign policy.[18] As Britain kept on withdrawing its overseas garrisons due to its shrinking domestic economy, Hong Kong was rendered indefensible against an overt attack, save with US support. Anglo-American relations involved complicated negotiations—not only over the protection of Hong Kong but also concerning the general relationship of the two Western powers and the extent to which Britain was willing to support US causes in international affairs. Mark argues that Hong Kong was put in a difficult position as a "reluctant Cold War Warrior"[19] in the Anglo-American alliance, as Britain had to juggle between the discordant commitments to maintaining friendly relations with China and avoiding ruptures in the American partnership.[20]

An allegorical reading of the production of *Ferry to Hong Kong* can deepen understanding of the underlying geopolitical narratives of Cold War Hong Kong. Gilbert agreed to make the film in the shadow of Britain's special

---

18. Chi-Kwan Mark, "A Reward for Good Behavior in the Cold War: Bargaining over the Defense of Hong Kong, 1949–1957," *International History Review* 22, no. 4 (December 2000): 837.

19. Chi-Kwan Mark, *Hong Kong and the Cold War: Anglo-American Relations, 1949–1957* (Oxford: Clarendon Press, 2004), 6.

20. For Hong Kong's unique Cold War experience in maintaining a pragmatic balance between China, Britain, and the United States, see Priscilla Roberts, "Cold War Hong Kong: Juggling Opposition Forces and Identities," in *Hong Kong in the Cold War*, eds. Roberts and John M. Carroll (Hong Kong: Hong Kong University Press, 2016), 26–59.

relationship with the United States, as embodied by Welles. The American filmmaker not only assumed an intimidating presence on the film set but also saw Hong Kong in Cold War terms as a city of political intrigue. Welles's remark reminds us of Edward Dmytryk's *Soldier of Fortune* (1955), in which Hong Kong was imagined as an "East Asian Casablanca," a safe transit place for refugees and escapees from Red China and a refuge for Hank Lee (Clark Gable), the US Navy deserter turned smuggler and pirate.[21] Welles observed that the frontier with China—"a collision of two worlds"—made the border one of the world's most dangerous and exciting contact zones. As a famed American filmmaker—and a tourist—Welles gained special permission from colonial authorities to visit the China border area. He appreciated it as "the only place in the world where Britain ha[d] a common frontier with a Communist country and the Union Jack [flew] opposite the Red Flag."[22]

When Welles alluded to Hong Kong as "the new Third Man territory," he could have been voicing British fears of Soviet aggression in central Europe as much as the infiltration of Chinese Communism in Hong Kong and Southeast Asia. By 1949, British policymakers propagated the view that Hong Kong was the "Berlin of the East"—the linchpin of the eastern front of the Cold War—where rival empires had converged in the British colony.[23] In September 1949, British foreign secretary Ernest Bevin told US secretary of state Dean Acheson that "Hong Kong was the rightwing bastion of the Southeast Asian front." If Hong Kong were to be lost to the Communists "the whole front might go."[24] Hong Kong governor Alexander Grantham,

---

21. For an analysis of *Soldier of Fortune*, see Thomas Y. T. Luk, "Hollywood's Hong Kong: Cold War Imagery and Urban Transformation in Edward Dmytryk's *Soldier of Fortune*," *Visual Anthropology*, no. 27 (2014): 138–48; Paul Cornelius and Douglas Rhein, "*Soldier of Fortune* and the Expatriate Adventure Film," *Quarterly Review of Film and Video*, March 7, 2021, https://doi.org/10.1080/10509208.2021.1891832.
22. Lewin, "Third Man in Hong Kong Part 1."
23. Christopher Sutton, *Britain's Cold War in Cyprus and Hong Kong: A Conflict of Empires* (Cham: Palgrave Macmillan, 2017), 152.
24. Quoted from Steven Hugh Lee, *Outposts of Empire: Korea, Vietnam, and the Origins of the Cold War in Asia, 1949–1954* (Montreal and Kingston: McGill-Queen's University Press, 1995), 18.

who served his term of office in 1947–1957, perceived the new Chinese Communist regime as "violently anti-Western, anti-British, and anti-Hong Kong."[25] The political referent of Berlin, which was the most prominent symbol of the Cold War in a split Europe, was revoked to reveal the precarious and "indefensible" situation of Hong Kong as seen by British eyes.

No less important than economic and strategic reasons, the prestige factor in international affairs was pivotal in driving the British government to maintain control and stability of Hong Kong. Prime minister Clement Attlee warned his cabinet that a failure to defend Hong Kong "would damage very seriously British prestige throughout the Far East and South-East Asia." Moreover, "The whole common front against communism in Siam, Burma and Malaya was likely to crumble unless the peoples of those countries were convinced of our determination and ability to resist this threat to Hong Kong."[26]

The wave of decolonization, which gained momentum after World War II, was restructuring the world order as well as bringing the British empire to an end. For Britain the Asian Cold War began with the declaration of the emergency in Malaya in 1948. A more remarkable Anglo-American joint venture was *The 7th Dawn* (1964), produced by Rank and also directed by Lewis Gilbert, about an American major (played by American star William Holden) who helped Malayan forces defend Malaya from the Japanese invasion during World War II. After the war, he was drawn into the British's counter-Communist operation in which he had to confront his former Chinese ally who joined the Communist rebel force. *The 7th Dawn* was released by United Artist in the United States, starring William Holden as one of the biggest box-office draws in Hollywood. The producer hoped that the film about British struggles in Cold War Asia "would help to make the American people as a whole more aware of the part Britain [was] playing

---

25. Alexander Grantham, *Via Ports: From Hong Kong to Hong Kong* (Hong Kong: Hong Kong University Press, 1965), 139.
26. Quoted from Lee, *Outposts of Empire*, 18.

against Communism in the Far East."[27] In addition, the US market was both commercially and diplomatically important for Rank to sell British war films abroad as well as to forge a united cultural front of British cinema and Hollywood in resistance to Communism.

The bargaining between British and American powers in international politics mirrors the nature of Anglo-American cooperation on the production of big-budget Cold War films as well as the delicate balance between this alliance and the anti-Communist coalition. Little known, however, is Welles's cinematic trajectory from *The Third Man*, a political noir that captured the paranoia of Vienna during the four-power occupation of Austria from 1945–1955, to *Ferry to Hong Kong*, in which his high-profile excursions and interviews in Hong Kong obviously showed us the condescending gaze of an American visitor who squarely exoticized as well as marginalized the local realities of the city. *The Third Man* symbolized the underlying tensions and vulnerability of an Anglo-American alliance against the Soviets.[28] The British could see themselves only as a diminishing empire caught between two monolithic powers, the United States and the Soviet Union.[29] With contextual and textual scrutiny, I argue that *A Ferry to Hong Kong* not only revealed to us the vulnerability of this alliance at the production stage but that the cinematic imaginary itself was rather self-critical of British nostalgia with an uncertainty of imperialist ambitions to preserve the colony as what

---

27. Harper and Porter, *British Cinema in the 1950s*, 45.
28. Reed later directed *The Man Between* (1953), which captured the demoralization of life in war-torn Berlin. But location shooting was barred in East Germany, and Reed could not replicate the visual virtuosity of *The Third Man*. Shaw, *British Cinema and the Cold War*, 70–74.
29. Lynette Carpenter, "'I Never Knew the Old Vienna': Cold War Politics and *The Third Man*," *Film Criticism* 3, no. 1 (1978): 28–29. After the defeat of Nazi Germany, Austria was subdivided into four occupation zones and jointly administered by the Soviet Union, the United States, the United Kingdom, and France. In the film, Vienna is a microcosm of international politics. The story is told from a British perspective (from producer Alexander Korda, director Carol Reed, and screenwriter Graham Green) and focuses on the precariousness of Vienna—and Europe—in the immediate postwar period: they could fall into the hands of the Soviet Union.

it used to be, in the period when Britain's global position was surpassed by the United States, and in which empire defense was reconfigured as imperial retreat.

Cold War Cinematic Memories

This article analyzes the representation of British imperial nostalgia in a spectacular British-produced picture and an Anglo-American joint venture, *Ferry to Hong Kong*, to throw light on the cinematic connections between Britain and Cold War Hong Kong as a complex historical situation, in which US forces were proactively undertaking cultural interventions, not least through the dominating presence of Hollywood. Imperial nostalgia is "associated with the loss of empire—that is, the decline of national grandeur and the international power politics connected to economic and political hegemony."[30] Postwar British cinema is a compelling vehicle to articulate the underlying tensions of the dissolution of the empire and the dwindling of Britain's world power.[31] In this study, *Ferry to Hong Kong* is taken as an imaginary battleground in Britain's ideological war against the forewarned Communist intrusions into the Crown colony.

A sentiment of loss of power or status underpins the nostalgic undertones in *Ferry to Hong Kong*. Affectively, the picture wrestles with the pain and rueful memory of a cherished past that was disappearing and embraces "a romance with one's own fantasy," where memories of the past are conditioned by the exigencies of the present crisis moment.[32] It clandestinely informs how the less powerful nation now has to come to terms with current

---

30. Patricia M. E. Lorcin, "Imperial Nostalgia; Colonial Nostalgia: Differences of Theory, Similarities of Practice?" *Historical Reflections* 39, no. 3 (2013): 107.
31. For a discussion of postwar British cinema and imperial nostalgia, see Stuart Ward, "Introduction," in *British Culture and the End of Empire*, ed. Ward (Manchester: Manchester University Press, 2001), 1–20.
32. Svetlana Boym, *The Future of Nostalgia* (New York: Basic Books, 2001), xiii–xvi.

political reality to account for the shifting international order in the postwar world. Drawing widely on primary sources as well as secondary accounts, including media responses upon the film's release and individual memoirs of production, this study probes into the evolution of the story of a sinking ferry from the real-life story behind the film—its fictionalization—to the events of its cinematization as a Cold War allegory.

The essay also addresses recent Cold War scholarship that has placed due emphasis on American foreign policies and strategies in containing the spread of Communism in Asia where Hong Kong played a key role. Sangjoon Lee has examined the Asia Foundation (TAF), a US nongovernmental agency that established Hong Kong as the primary center of media production—especially newspapers, magazines, and movies—to promote non-Communist and counter-Communist materials for overseas Chinese audiences in Southeast Asia. The significance of Hong Kong was fundamentally attributed to its geographical, political, and economic weight among overseas Chinese communities.[33] Poshek Fu also points out that TAF was deeply involved in the cultural warfare in Hong Kong, while motion pictures constituted a major instrument of propaganda—namely, the use of moving images to contain Communist influences in overseas Chinese communities.[34] Alongside TAF, United States Information Services (USIS) played a fairly important role in producing Chinese-language pictures in Hong Kong and Southeast Asia to counter the cultural and political impacts of pro-Communist cinema in the regions.[35]

33. Sangjoon Lee, "Creating an Anti-Communist Motion Picture Producers' Network in Asia: The Asia Foundation, Asia Pictures, and the Korean Motion Picture Cultural Association," *Historical Journal of Film, Radio and Television* 37, no. 3 (2017): 517–38.

34. Poshek Fu, "Entertainment and Propaganda: Hong Kong Cinema and Asia's Cold War," in *The Cold War and Asian Cinemas*, eds. Fu and Man-Fung Yip (New York: Routledge, 2019), 238–62.

35. For the role of USIS in propagating anti-Communist literature, see Kenny K. K. Ng, "Soft-Boiled Anti-Communist Romance at the Crossroads of Hong Kong, China, and Southeast Asia," in *Chineseness and the Cold War: Contested Cultures and Diaspora in Southeast Asia and Hong Kong*, eds. Jeremy E. Taylor and Lanjun Xu (New York: Routledge, 2021), 94–109.

While the United States took Hong Kong as its major propaganda hub and was proactive in engaging anti-Communist politics in Asia, British officials sought to prevent overt propaganda efforts by both the United States and China; hence, they pursued a pragmatic tactic to sustain the "'careful fiction' of Hong Kong's neutrality in the Cold War."[36] In domestic politics, the Hong Kong government tried to maintain political neutrality and a nonconfrontational approach vis-a-vis the threatening presence of Chinese Communists across the border. The colony was a "Cold War grey zone" where "certain communist activities were tolerated but rigidly confined by the colonial legal frame."[37] Britain preferred appeasement to confrontation in order to maintain the status quo of Hong Kong.

In this light, the British never deliberately released their own war films or propaganda works in Hong Kong. Even though the colony was formally under British rule, they did not seek to advance their political cause or interest in this fashion.[38] A notable example is Michael Anderson's *Yangtse Incident: The Story of HMS Amethyst* (aka *Battle Hell*) (1957). This was a British-made war movie that hailed British naval valor against the Communist Chinese army's attack on the Yangtze in 1949. There is no obvious local release record of the film in Hong Kong.[39]

On April 19, 1949, the British frigate HMS *Amethyst* was ordered up the Yangtze River to act as a guardship for the British embassy in Nanjing

---

36. Charles Leary, "The Most Careful Arrangements for a Careful Fiction: A Short History of *Asia Pictures*," *Inter-Asia Cultural Studies* 13, no. 4 (2012): 548.
37. Yan Lu, "Limits to Propaganda: Hong Kong's Leftist Media in the Cold War and Beyond," in *The Cold War in Asia: The Battle for Hearts and Minds*, ed. Zheng Yangwen, Hong Liu, and Michael Szonyi (Leiden: Brill, 2020), 95.
38. Kenny K. K. Ng, "Inhibition vs. Exhibition: Political Censorship of Chinese and Foreign Cinemas in Postwar Hong Kong," *Journal of Chinese Cinemas* 2, no.1 (2008): 26.
39. There was no clear evidence of *Yangtse Incident* ever being shown in Hong Kong according to available records. A report from the *South China Morning Post* mentioned: "Unfortunately this film is not likely to be seen as newer films are already available for preview in the Colony." See Jean Gordon, "Around the Cinemas: Most Popular Films & Stars of Last Year," *South China Morning Post*, April 16, 1958, 4. A report a year later stated that the film was not yet released. Jean Gordon, "Around the Cinemas: Cooper and Heston Dive in Sunken Hold," *South China Morning Post*, August 20, 1959, 4.

during the Chinese civil war. The ship came under fire from Communist artillery batteries on the northern bank of the river, and it ran aground. After being held in custody for three months by the Communists, whose demand for a British apology was firmly refused, the ship, under the command of Lieutenant Commander John Kerans, fled 167 kilometers (103.7 miles) in the dark to rejoin the British fleet in Shanghai. During the incident seventeen members of the crew were killed and ten wounded.[40] The frigate arrived in Hong Kong in August under a glare of publicity from the world's press.

With the intention of reconstructing an official British version of the ship's three-month ordeal, *Yangtse Incident* warranted official support from the foreign office, the navy, and the veterans who went through the actual warfare. The national production gave meticulous attention to historical and technical details. John Kerans himself was the film's technical advisor, and the real HMS *Amethyst* was used in the film—in fact, it was even more badly damaged during the filming than in the real warfare. While the British propagandistic war film was "a smash for [the] home market" in Britain in 1957,[41] it failed to charm American moviegoers or French critics at the Cannes Film Festival. The film received more attention at home than abroad.[42] Foreign viewers, even those hailing from Britain's close allies, greeted it with lackluster enthusiasm; they were not interested in the British national pride it conveyed. Even British commentator Derek Hill was annoyed at its nostalgia: "The British cinema seems to be looking backward because it lacks the courage or honesty to look forward. And nostalgia over past military exploits is even more dangerous than nervousness about the future."[43]

---

40. Since the late nineteenth century, the Royal Navy had protected British interests and international riverine commerce in China. For an account of the *Amethyst* incident, see Edwyn Gray, "The Amethyst Affair: Siege on the Yangtze," *Military History* 16, no. 1 (April 1999): 58–64.
41. Myro, "Film Reviews: *Yangtse Incident* (British)," *Variety*, April 10, 1957, http://archive.org/details/sim_variety_1957-04-10_206_6.
42. Edouard Laurot, "Cannes Film Festival," *Film Culture*, October 1957, 8–9; H.H.T., "*Battle Hell*: Richard Todd Stars in British Import," *New York Times*, August 22, 1957.
43. Derek Hill, "Down-At-Heel Dignity–At Your Cinema," *Amateur Cine World*, June 1957, 160–62.

Yangtse Incident emerged in an age of anxiety when the empire was falling apart. The war film did not fare well enough at the box office to recover its production costs. It premiered in 1957 in Britain, two years before *A Ferry to Hong Kong* was screened in Hong Kong. Indeed, Gilbert was associated with a dozen war films of the 1950s made to commemorate war heroes in "the people's war."[44] Gilbert acknowledged that the war film genre arose because "after the war Britain was a very tired nation." Already in the mid-1950s, war films were needed as "a kind of ego boost, a nostalgia for a time when Britain was great," since economically the country was already falling behind the defeated nations of Germany and Japan.[45]

Major British studios like Rank were as much enmeshed in the ideological battle against Communist subversion as they were eager to vindicate British values. The British government between 1945 and 1965 sought to "integrate the cinema within the anti-communist and anti-Soviet propaganda campaign"; films were seen as "potentially active producers of political and ideological meanings."[46] Before producing *A Ferry to Hong Kong*, Rank specialized in "colonial war" films, which articulated British memories of World War II and the uprisings in its colonies.[47] Nevertheless, public sentiment changed in Britain with the Suez Canal debacle and the ensuing demise of British and French colonial power.[48] The humiliation at Suez meant to the world that the empire was no longer a source of political strength for

---

44. John Ramsden, "Refocusing 'The People's War': British War Films of the 1950s," *Journal of Contemporary History* 33, no. 1 (January 1998): 35–63.
45. Ramsden, "Refocusing 'The People's War,'" 59.
46. Shaw, *British Cinema and the Cold War*, 3.
47. These colonial war films were *The Planter's Wife* (1952), regarding the Communist insurgency in Malaya (1948–1960); *Simba* (1955), regarding the Mau Mau uprising in Kenya (1952–1960); *Windom's Way* (1957), set in Malaya; and *North West Frontier* (1959), set in India. See Harper and Porter, *British Cinema in the 1950s*, 44–46.
48. Kazuo Ishiguro's *The Remains of the Day* (1989), a novel about a fading English aristocracy and the regrets of lost love, takes as its starting point the historical setting of July 1956, which coincides with the beginning of the Suez crisis.

Britain, which had already lost its leading position to the United States and the Soviet Union as the new global superpowers.[49]

By 1958, British viewers were fond of grimmer and more sober remembrances of the war. Two large-budget productions, Columbia's *The Bridge on the River Kwai* (dir. David Lean, 1957) (starring William Holden) and Ealing's *Dunkirk* (dir. Leslie Norman, 1958), topped the British box-office polls.[50] These cinematic representations and audience responses could have been telling signs that "the British psyche under stress was vulnerable, volatile, and potentially unreliable."[51] These popular war films played a role in restoring Britain's prestige in the context of British isolationism and anxiety and conveying the enduring theme of individuals fighting against impossible odds in the madness of war.[52]

But Gilbert's film did not follow the popular success of *The Bridge on the River Kwai*. The British American war epic and adventure story mixed personal conflicts and ordeals with heroism, valor, and anti-war ideals. *Ferry to Hong Kong*, instead, was informed by the allegorical storytelling of the competing interests of Britain and America in screening Hong Kong to the (Western) world. Hollywood reacted favorably to American foreign policy orientations to produce imaginary pictures of America's Cold War confrontations with

---

49. In 1956, Egypt's President Gamal intended to nationalize the management of the Suez Canal. Both shareholders of the canal company, Britain and France were irritated by Egypt's decision. The two European powers launched an invasion of Egypt. The US president Dwight D. Eisenhower disapproved of the offensive and demanded a military withdrawal as the attack was a reproduction of the old Western colonization. See G. C. Peden, "Suez and Britain's Decline as a World Power," *Historical Journal* 55, no. 4 (2012): 1073–96.

50. *The Bridge on the River Kwai* illustrates the lunacy of war and the moral dilemmas of the British and American POWs who are coerced to build a railway bridge across the river Kwai for their Japanese captors in occupied Burma. *Dunkirk* is the cinematic reconstruction of the events that took place between May 26 and June 4, 1940, when approximately 336,000 British, French, and Belgian troops were evacuated from the beaches of Dunkirk in northern France by the combined efforts of naval and civilian crews. Dunkirk is considered the point at which World War II began.

51. Harper and Porter, *British Cinema in the 1950s*, 136.

52. Penny Summerfield, "*Dunkirk* and the Popular Memory of Britain at War, 1940–58," *Journal of Contemporary History* 45, no. 4 (October 2010): 788–811.

Communist China and heroic adventures set in Hong Kong, such as *Soldier of Fortune*, where the British were nominally in power. Hollywood's propaganda and soft power often counted on films that portrayed the Communists as a dark and barbaric force that infiltrated and threatened to destroy America.[53] *Ferry to Hong Kong* tended to be more subtle and subdued. It conveyed sentiments of nostalgia and loss by portraying a descendant of the doomed Austrian and British empires as an exile condemned to a drifting shipboard life. It was a cinematic exemplar of British soft propaganda, a self-defensive but self-defeating eulogy of British national influence on Asian Cold War fronts. Yet, in competing with Hollywood, the British film was locked within the Western cinematographic cliché and Cold War Orientalism.

Cold War Coproduction and Orientalism

The long-forgotten event represented in *Ferry to Hong Kong* and British-American joint endeavors have been a subject of Cold War film studies in East Asia. Stephanie De Boer has examined Hong Kong-Japan coproductions made from the 1950s to the 1970s in the shadow of Japanese imperial occupation of the region and its postwar media legacy.[54] Cinematic coproductions, as Erica Ka-yan Poon argues, was an intensified site of cultural battles for prestige in Cold War Asia.[55] The event represented in *Ferry to Hong Kong* makes for a cogent case study of British film production liaising and vying with Hollywood and underlies both imperial nostalgia and colonial geopolitics in the 1950s.

---

53. See Russell E. Shain, "Hollywood's Cold War," *Journal of Popular Film* 3, no. 4 (1974): 334–50; Jeffrey Richards, *China and the Chinese in Popular Film: From Fu Manchu to Charlie Chan* (London: I. B. Tauris, 2017), 177–95.

54. Stephanie DeBoer, *Coproducing Asia: Locating Japanese-Chinese Regional Film and Media* (Minneapolis: University of Minnesota Press, 2014).

55. Erica Ka-yan Poon, "Southeast Asian Film Festival: The Site of the Cold War Cultural Struggle," *Journal of Chinese Cinemas* 13, no.1 (2019): 76–92.

The production of *Ferry to Hong Kong* cost half a million pounds (roughly $1,400,000) and marked Rank's most ambitious project to date. Rank sought to compete with Hollywood studios by making increasingly lavish international coproductions with CinemaScope color photography to be sold at home and abroad. In the film's early stages of planning, Rank's chairman John Davis decided that "an exotic title, and American leading man and a continental director" were needed to make a successful international picture.[56] The declining turnout at domestic theaters amid the competition from television,[57] as well as the dominating presence of Hollywood, forced Rank to shift its strategy to produce only expensive blockbusters that "had international entertainment appeal" and could be "vigorously sold in foreign markets."[58]

In terms of technology, *Ferry to Hong Kong* was the first of Rank's feature films that adopted the US wide-screen format of CinemaScope. In Hong Kong, the production team converted a ferry into a paddle steamer specifically for moviemaking and hired local labor to build a studio stage and a crane for the CinemaScope camera.[59] The on-location shooting had been widely practiced by Rank in (former) British colonies and the Commonwealth countries. The studio took full advantage of the newest Eastman Color innovation and lightweight cameras to make a range of films that exploited exotic locations.[60] Made with the cosmopolitan conformity of international production, and invested with clichés of Orientalism (Chinese pirates, seductive dancers, thugs and gangsters, and devoted lovers in an underdeveloped Asian land), the film attempted to tap into the fascination of international audiences with the exotic and the spectacular in the 1950s.

---

56. Quoted from Geoffrey Macnab, *J. Arthur Rank and the British Film Industry* (London: Routledge, 1993), 225
57. Stephen Watts, "Rank Theatre Chain, Production List Reduced," *New York Times*, October 26, 1958.
58. Harper and Porter, *British Cinema in the 1950s*, 52.
59. A. H. Weiler, "Endless Trip," *New York Times*, September 28, 1958.
60. Harper and Porter, *British Cinema in the 1950s*, 208.

In Hollywood's most iconic Hong Kong films, *Love Is a Many-Splendored Thing* and *The World of Suzie Wong*, Hong Kong was generically featured as an exotic tropical locale where everyday life and local realities were sidelined to give way to exoticism and desire.[61] The Cold War city was positioned as a place of refuge and a site of intercultural romance, where "a postwar American identity can be defined against an emerging Asian communism and the decay of European colonialism."[62] *Love Is a Many-Splendored Thing* opens with an astounding bird's-eye view of Hong Kong Island and the harbor. The camera flies over and runs parallel to the waterfront until it cuts to a scene of urban streets and chaotic traffic, following an ambulance along Queen's Road. Alfred Newman's themed music sets the film's affective tone with a sense of aloofness, whereas the lovers' rendezvous occurs on an idyllic, grassy hilltop (shot in rural California), which is alienated from the urban density and social reality of Hong Kong. By comparison, *The World of Suzie Wong* has more engagement with Hong Kong's urban locality, with scenes of the street markets and bars in Wanchai. The white male protagonist ventures into the shanty towns and witnesses the awful living conditions of Chinese refugees. The location shots of urban reality turn the exotic locale into a dystopic "lawless wilderness."[63]

Recalling Hollywood's dominant aesthetic temperament of location-shooting, the opening montage sequence of *Ferry to Hong Kong* follows the nomadic protagonist Conrad walking from the waterfront of western Kowloon to overcrowded market streets on Hong Kong Island, transitioning from day to night when Conrad passes by the Dragon nightclub with its clichéd oriental flavor (fig. 4).

---

61. Wendy Gan, "Tropical Hong Kong: Narratives of Absence and Presence in Hollywood and Hong Kong Films of the 1950s and 1960s," *Singapore Journal of Tropical Geography* 29 (2008): 8–23.

62. Gina Marchetti, *Romance and the "Yellow Peril": Race, Sex, and Discursive Strategies in Hollywood Fiction* (Berkeley: University California Press, 1993), 110.

63. Elleke Boehmer, *Colonial and Postcolonial Literature: Migrant Metaphors* (Oxford: Oxford University Press, 1995), 44.

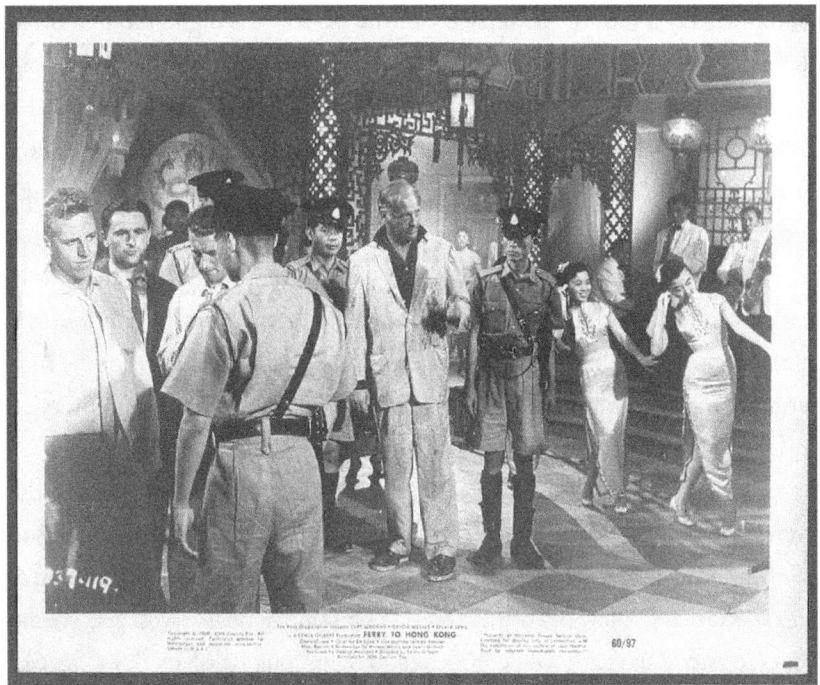

Figure 4: The Oriental nightclub in *Ferry to Hong Kong.* © IMDb.
Source: 20th Century Fox, *Ferry to Hong Kong*, 1960, printed still, IMDb,
https://www.imdb.com/title/tt0052799/mediaviewer/rm572475904/.

The sequence is preceded by the juxtaposition of a modern steamship and a junk in the harbor as a stock phrase of East meets West. It cuts to a floating sampan as we see Conrad wake up and walk off to the shore. Dressed in a soiled suit and sleeping on the boat, covered by newspapers for warmth, Conrad is presented as a man without a home, a tramp on an outdoor adventure. The wide-screen format permits more lateral camera movement to capture Conrad's walk among the boat people across the fishing shelter, showing the breadth of the physical space of the working-class neighborhood. His wandering is accompanied by a soundtrack of Nanyin (Southern tunes) singing, the local Cantonese vernacular music performed by blind singers to entertain lower-class people. The establishing sequence is followed by a shot of the sunset over the sea and the nightclub in a bright neon light, framed and illuminated by the CinemaScope and Eastman Color to transform the shore footage into an oriental spectacle. "Photographed in color entirely in the Hong Kong area," a critic noted, "the Twentieth Century-Fox release is drenched in exotic atmosphere. It fairly simmers as the opening credits come on."[64]

The Orientalism of *Ferry to Hong Kong* extended beyond the movie screen to real-life cinematic circles and social gatherings in London. The film was promoted in its British homeland as a blockbuster, "one of the most ambitious British films for years."[65] It received a spectacular world premiere in London attended by one thousand guests on July 2, 1959 (fig. 5).[66]

Popular media compared the extravaganza to the producer Mike Todd's grand premiere party for *Around the World in 80 Days* (dir. Michael Anderson, 1956). The glamorous oriental-themed party greeted the guests with fireworks and spectacles.[67] A few celebrities traveled on sampan boats across the Thames River to attend the party while Jurgens and Syms took rickshaw rides (fig. 6).[68]

In a cinematic setting glitzed up by Chinese lanterns and dragon icons, they enjoyed a sumptuous buffet of exotic Chinese dishes, such as honeyed grasshoppers, served with chopsticks and delivered by ethnic Chinese girls in *cheongsams* (fig. 7).

The guests were entertained with variety shows featuring Chinese acrobats, dragon-dance performers, jugglers, fortune-tellers, and an orchestra band.[69]

---

64. Howard Thompson, "*Ferry to Hong Kong* Stars Orson Welles," *New York Times*, April 27, 1961, 27.
65. Reuters, "It Rivalled Even Mike Todd's Party! *Ferry to Hongkong* Given Exotic Send-off in London," *China Mail*, July 3, 1959.
66. Reuters, "Film Premiere," *Daily Telegraph and Morning Post*, July 3, 1959.
67. "Premiere of *Ferry to Hongkong*: Chinese Setting to U.K. Party," *South China Morning Post*, July 4, 1959, 9.
68. Paul Tanfield, "Is There at the Night of Thousand Lanterns: This Was Mike Todd Stuff! Sampans Make a Splash on the Thames," *Daily Mail*, July 3, 1959.
69. James Fazakerley, "'Ferry' Wakes up London with a Bang and Makes £20,000 for Charity," *South China Morning Post*, July 9, 1959, 9; Gilbert, *All My Flashbacks*, 195–96.

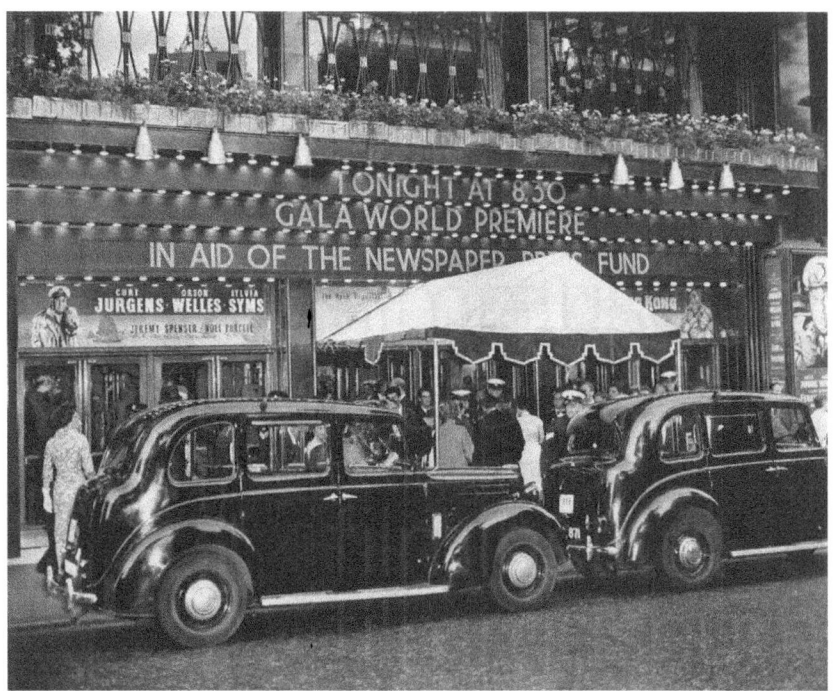

Figure 5: The oriental-themed gala world premiere of *Ferry to Hong Kong* on July 2, 1959, at 8:30 p.m., at the Odeon Theatre Leicester Square, London, with guests arriving in *cheongsam*. Reproduced with the permission of Deutsches Filminstitut & Filmmuseum.
Source: Ferry to Hong Kong (1959) Premierenfoto 3, 1959, photograph. Curt Jürgens: The Bequest, Deutsches Filminstitut & Filmmuseum, https://curdjuergens.deutsches-filminstitut.de/nachlass/ferry-to-hong-kong-1959-premierenfoto-3/. Reproduced by permission of Deutsches Filminstitut & Filmmuseum.

In Hong Kong, the film stars had no qualms about showing their (self-) Orientalist appreciation of the Chinese lifestyle in the colony to Anglophone media, which followed their sightseeing tours when they were not filming. Jurgens and Syms put on Chinese clothing, collected Chinese antiques, and tried outlandish Cantonese foods like snake soups and hot pots.[70] Jurgens

---

70. Lewin, "Third Man in Hong Kong Part 2."

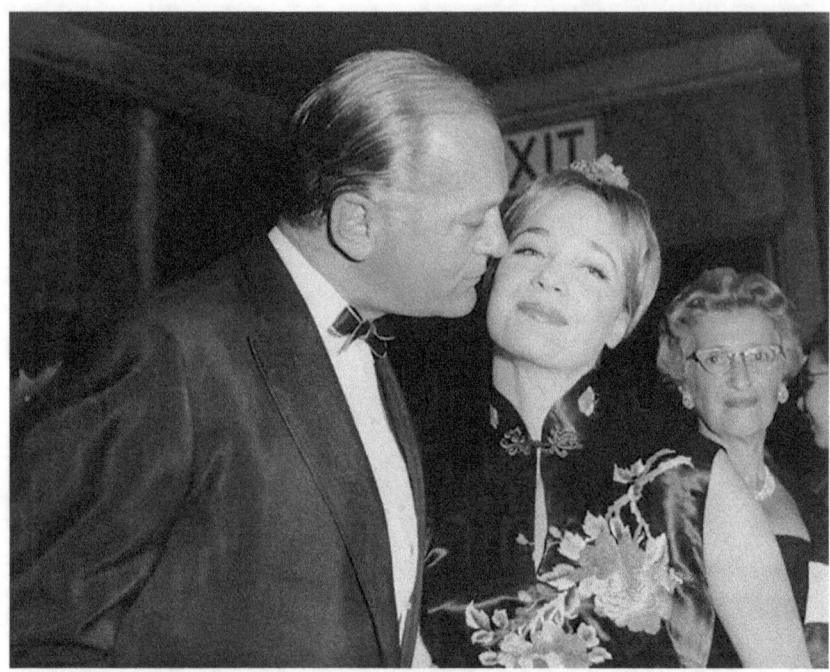

Figure 6: Curt Jurgens and Sylvia Syms at the UK premiere of *Ferry to Hong Kong*. Reproduced with the permission of Deutsches Filminstitut & Filmmuseum.
Source: Ferry to Hong Kong (1959) Premierenfoto 1, 1959, photograph. Curt Jürgens: The Bequest, Deutsches Filminstitut & Filmmuseum, https://curdjuer gens.deutsches-filminstitut.de/nachlass/ferry-to-hong-kong-1959-premierenfo to-1/. Reproduced by permission of Deutsches Filminstitut & Filmmuseum.

and his spouse Simone Bicheron visited the Precious Lotus Temple on Lantau Island, expressing amazement at the sedan chair ride as they were carried uphill by Chinese coolies (fig. 8).[71]

The stars marveled at their tourist excursions as they experienced and imagined how upper-class expatriates lived in Hong Kong (fig. 10).[72]

---

71. Lewin, "Third Man in Hong Kong Part 3: Rare Survival: A Genuine Flamboyant Star!" *China Mail*, March 4, 1959.
72. By comparison, the American Welles showcased an Orientalism that was more self-conscious and oriented toward the grassroots. He drank "spell water" to wish good fortune

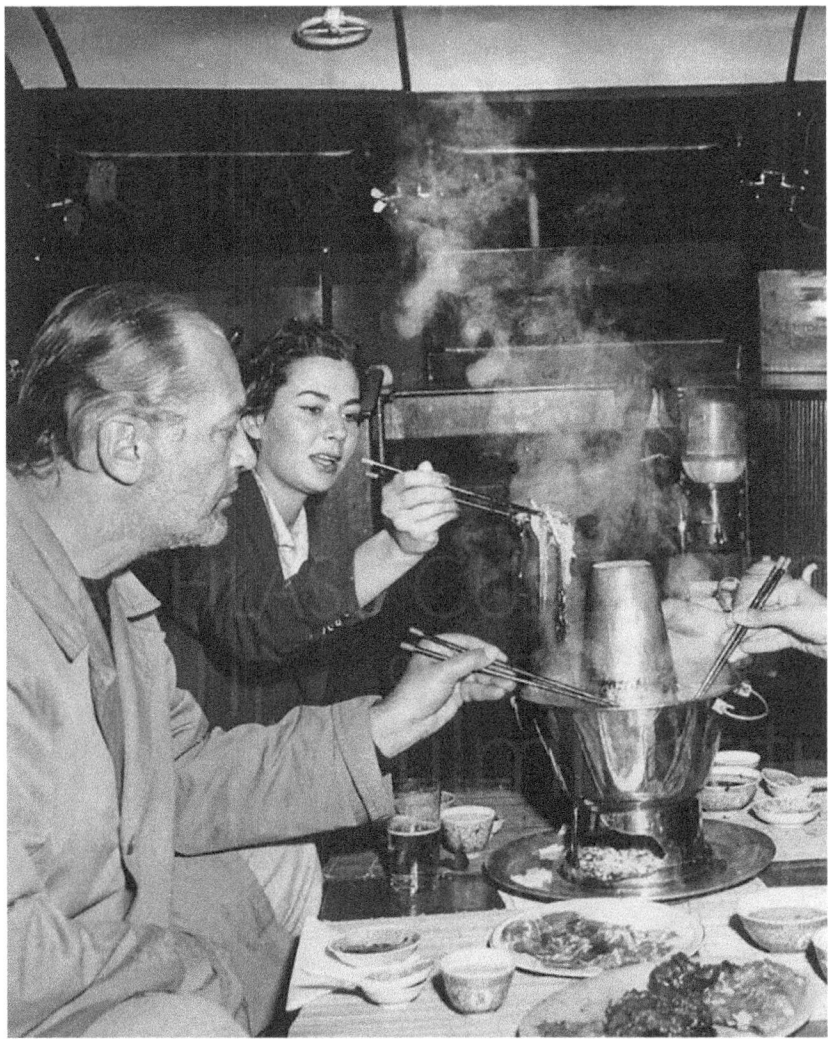

Figure 7: Curt Jurgens and his spouse Simone Bicheron enjoying Cantonese hot pot in Hong Kong. Reproduced with the permission of Deutsches
Filminstitut & Filmmuseum.
Source: Ferry to Hong Kong (1959) Curd und Simone privat 8, 1959, photograph. Curt Jürgens: The Bequest, Deutsches Filminstitut & Filmmuseum,
https://curdjuergens.deutsches-filminstitut.de/nachlass/ferry-to-hong-kong-
1959-curd-und-simone-privat-8/. Reproduced by permission of Deutsches
Filminstitut & Filmmuseum.

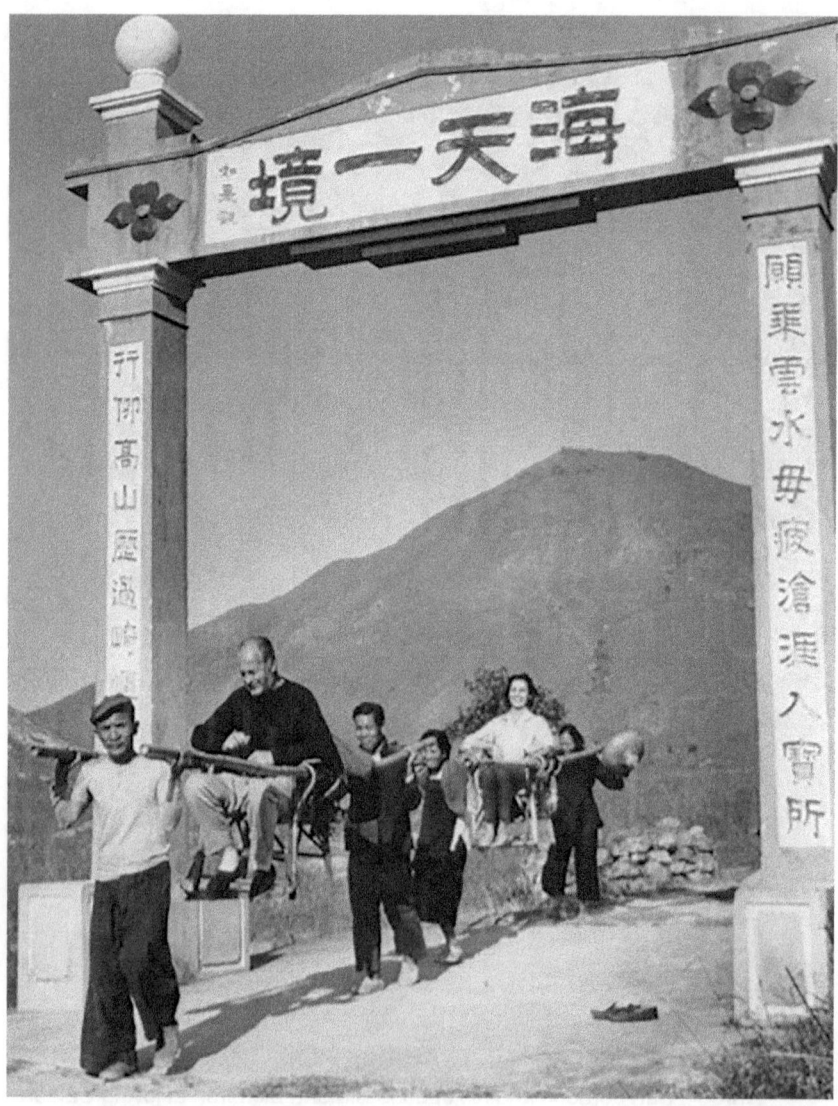

Figure 8: Curt Jurgens and Simone Bicheron on sedan chairs in Hong Kong. Reproduced with the permission of Deutsches Filminstitut & Filmmuseum.

*Source: Ferry to Hong Kong (*1959) Curd und Simone Privat 13, circa 1959, photograph. Curt Jürgens: The Bequest, Deutsches Filminstitut & Filmmuseum, https://curdjuergens.deutsches-filminstitut.de/nachlass/ferry-to-hong-kong-1959-curd-und-simone-privat-13/. Reproduced by permission of Deutsches Filminstitut & Filmmuseum.

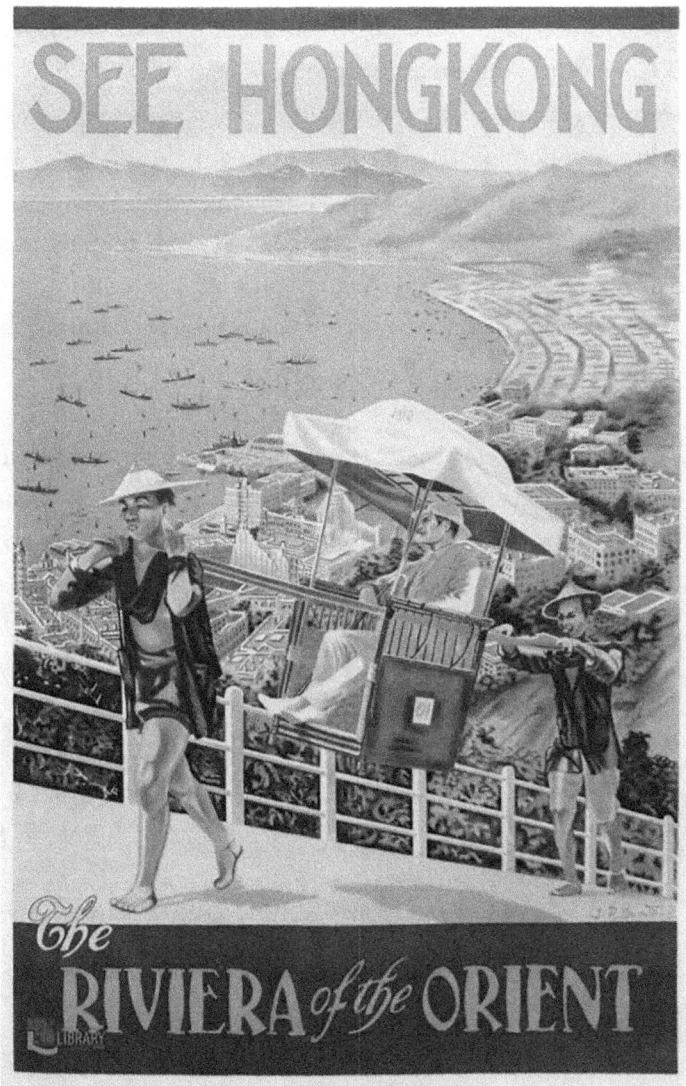

Figure 9: A 1935 poster commissioned by the Hong Kong Tourism Association. Poster by S. D. Panaiotaky. Image courtesy of Hong Kong Baptist
University Library.
Source: S. D. Panaiotaky, *See HongKong, the Riviera of the Orient*, circa
1935, poster, color, 105 cm. x 67 cm. on cloth board. Hong Kong Travel
Association, Hong Kong, in Hong Kong Baptist University Library Art Collections, https://bcc.lib.hkbu.edu.hk/artcollection/91512504h757t2/. Image
courtesy of Hong Kong Baptist University Library.

The zealous coverage in the English media of the stars' tourist activities revealed Cold War Hong Kong's status not only as a favorite film location for Western—predominantly Hollywood—moviemakers but also as a popular tourist desination in Asia. This was particularly true of Americans, who made up the largest national group of foreign visitors. Like a host of foreigner filmmakers before him who made trips to the city to shoot travelogue footage of their exotic Orient,[73] Welles took the opportunity to privately obtain movie footage of the city and documented local customs like Chinese weddings and funerals.[74] Postwar Hong Kong saw a ten-fold surge in tourists from the 1950s to 1960s, making the city the second most popular destination in Asia, after Japan.[75] It was American businesmen and executives who played an important role in promoting Hong Kong as Asia's tourist hub.[76] From the Korean War (1950–1953) to the escalation of the Vietnam War in the 1960s, the US military and the Seventh Fleet also used Hong Kong as a liberty port as well as a "rest and recreation" (R&R) center for thousands of American soliders and sailors.[77] American tourism and

---

for the crew according to Chinese superstition. He knew that he would not find chop suey in Hong Kong as the cuisine was created by diasporic Chinese railway workers in North America for American consumption. See Lewin, "Third Man in Hong Kong Part 4: The Fluid Face of Orson Welles," *China Mail*, March 5, 1959.

73. Foreign travelers and pioneers to Hong Kong had recorded city life with the new invention of the movie camera since the last nineteenth century. The oldest reel of travelogue documentaries preserved is the Edison Shorts, made in 1898 by the Edison Company. See "Transcending Space and Time: Early Cinematic Experience of Hong Kong (Early Motion Pictures of Hong Kong)," Hong Kong Film Archive, March 21–June 22, 2014, https://www.hkmemory.hk/MHK/collections/ECExperience/about/index.html.

74. "Departure of Orson Welles: Hope to Come Back to Shoot Own Picture," *South China Morning Post*, February 28, 1959, 12; David Lewin, "Third Man in Hong Kong Part 2: Behind the Ebullience: Welles Confesses a Lonely Bitterness," *China Mail*, March 3, 1959.

75. Chi-Kwan Mark, "Hong Kong as an International Tourism Space: The Politics of American Tourism in the 1960s," in *Hong Kong in the Cold War*, 163.

76. John Carroll, "'The Metropolis of the Far East': Tourism and Recovery in Postwar Hong Kong," Transnational Hong Kong History Seminar Series, NTU History and Hong Kong Research Hub, Nanyang Technological University of Singapore, February 10, 2022.

77. Chi-Kwan Mark, "Vietnam War Tourists: US Naval Visits to Hong Kong and British-American-Chinese Relations, 1965–1968," *Cold War History* 10, no. 1 (February 2010): 1–28.

Hollywood blockbusters were thus inextricably related to forging the spectacle of Hong Kong to be imagined and consumed by global visitors and by viewers on the screen.

In addition, Asia had seen rapid economic growth in the early 1950s and become an expanding market for international motion pictures: "The natives are particularly interested in Hollywood films."[78] An exemplary popular success was *The World of Suzie Wong*, adapted by American producer Ray Stark from a best-selling novel by British author Richard Mason. Both the fictional and cinematic storytelling infamously propagated foreign images of Hong Kong/Wanchai as the exotic "world of Suzie Wong," with girlie bars, dance halls, brothels, and restaurants catering to foreign visitors and sailors.[79]

Released only one year apart, both *The World of Suzie Wong* and *A Ferry to Hong Kong* shared Hollywood's narrative interest and commercial drive to popularize the tourist gaze on the screen. With the advantage of on-the-spot shooting, the movies could visualize the picturesque environments in color and sound, evoking an ethnic charm of working-class crowds hustling and bustling in street markets and the boat people living on waterfront shelters. But the British film could not compare with the success of Hollywood's tale of Suzie Wong, which fueled the popular American fantasy of the "white knight," a role assumed by a Caucasian male lover to safeguard his Asian girl.[80] Sylvia Syms, who starred in both films, plays a wealthy upper-class British banker's daughter, Kay O'Neill, in *The World of Suzie Wong*. A condescending British lady who despises the intimate relationship between her American painter friend Robert Lomax (William Holden) and his poor Asian lover, Suzie (Nancy Kwan), she acts as a romantic foil. Not only does the American hero save Suzie from the debris of war in poverty-stricken Hong Kong/Asia, but he also releases her from the grip of British violence and racism by physically fighting a British sailor who harasses Suzie and

---

78. Thomas M. Pryor, "British Film Aide Sees Asia Market," *New York Times*, November 1, 1952, 17.
79. Mark, "Hong Kong as an International Tourism Space," 166.
80. Marchetti, *Romance and the "Yellow Peril,"* 109–24.

rejecting the temptations of O'Neill's patronage. Robert shows his American distaste for British hierarchy and for business. By portraying an interracial love affair, the Hollywood film hides its own racism and exoticism.

In *Cold War Orientalism: Asia in the Middlebrow Imagination, 1945–1961*, Christina Klein explains how American middlebrow cultural products such as films, stage plays, magazines, and travel writings translated US foreign policy and shaped American audiences' belief in the role of the United States in Cold War Asia as the defender of democracy and freedom regardless of race and class.[81] Klein argues that American middlebrow fiction and films functioned to reshape its Orientalism to propel the myth of America-Asia bonding in a consolidation of the Free World in Asia.[82] Similarly, *The World of Suzie Wong* was ideologically important for disseminating American values of freedom to audiences.

In *Ferry to Hong Kong*, however, the romance of the Caucasian lovers Ferrers and Conrad could not excite the Orientalist imagination of American viewers. Rather, it was the understated lead Jurgens who, as the fugitive, gave the film reflexive depth with respect to British self-image and its identity crisis. The film changes the novel's American protagonist Clarry Mercer to British Austrian antihero Conrad. The name no doubt recalls Joseph Conrad's characters and their misadventures in the "primitive" lands of British colonies. In *Heart of Darkness* (1899), Marlow's mythical journey up the Congo River relates to his quest for the self; he is caught between the "civilized" and "uncivilized" worlds. In the film, Conrad in his uninvited excursions is met with contempt and hatred by his civilized British compatriots on board. A snobbish old lady in Victorian dress casts a derogatory look at Conrad, who cannot pay the travel fare at the entrance, when boarding the upper deck. Captain Hart despises him as a drunk and brawler. Calling Conrad "human refuse," Hart throws him down to the lower deck with

---

81. Christina Klein, *Cold War Orientalism: Asia in the Middlebrow Imagination, 1945–1961* (Berkeley: University of California Press, 2003).
82. Klein, *Cold War Orientalism*, 146.

the Chinese passengers. The outcast's experience offers viewers a glimpse of the inequality of different social classes in colonial society as reflected on the boat. By colonial law, Conrad is an "undesirable alien" expelled by the British Hong Kong police and must obey official instructions that refuse his entry to Hong Kong and Macau.

The film is inescapably stuck in a double bind between its (self-)mockery of British class society and imperialism and its Cold War Orientalism. Like Marlow's voyage of self-discovery in *Heart of Darkness*,[83] Conrad's exilic journey in *Ferry to Hong Kong* illuminates the truth through the eyes of the underdog hero: imperialism is only a shadowy presence. But the ideological battle for hearts and minds dictates the film's anti-Communist storytelling. The climax is freighted with thrilling action when storms and rains thrust the vessel into the darkness of the South China Sea, where pirates board the ship to rob the wealthy passengers. These uncivilized natives are none other than the Chinese Communists, who plunder the ferry. Conrad shines in his chivalric valor by heroically ridding the stranded ferry of Chinese pirates to save the passengers. Aided by Captain Hart and the crew, Conrad overcomes the pirates and comes close to steering the storm-battered ferry back to port, but it sinks in Hong Kong harbor (fig. 10).

Hong Kong actor Roy Chiao plays the Americanized Chinese Johnny Sing-Up, a black-leather-jacket and blue-jean-clad Elvis-styled rogue who teams up with Master Yen, the bald-headed, brutal pirate leader. Johnny describes his partner to the passengers as "a vulgar man of action," but "he is civilized like me" (fig. 11).

This understated line might remind us mockingly of how similar the violent looting by uncivilized pirates really is to the taking of territory by civilized colonial masters.

Chiao's participation in the Anglo-American blockbuster attracted much Chinese media attention. In an interview published in *International Screen*,

---

83. For a critique of imperialism in *Heart of Darkness*, see Edward Said, *Culture and Imperialism* (New York: Vintage Books, 2012), 19–30.

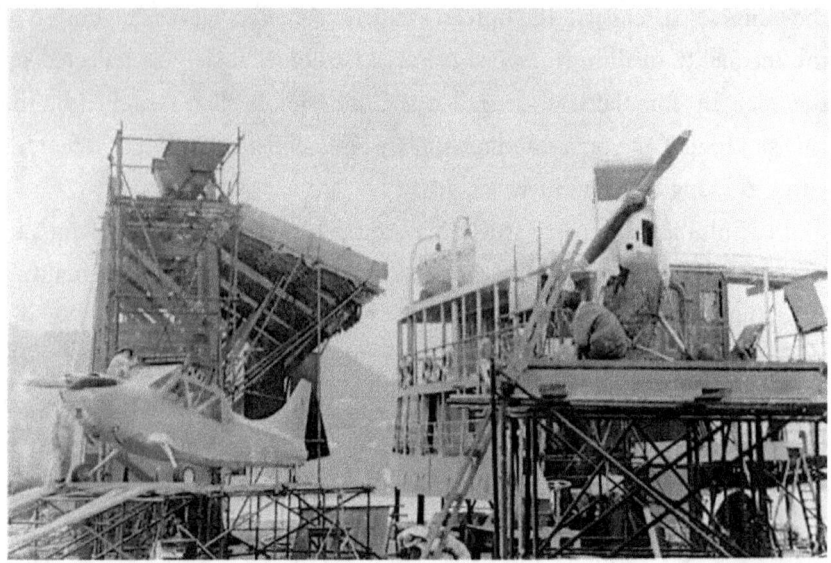

Figure 10: The storm scenes were filmed at Aberdeen using a mock-up of the ferry and two airplanes from the Far East Flying Training School at Kai Tak to provide the wind and water chutes to simulate waves. Photo by H. K. Watt, principal of the Far East Flying Training School. Image courtesy of the Hong Kong Historical Aircraft Association.
Source: H. K. Watt, *Ferry to Hong Kong Film*, 1958, photograph. Archives of Hong Kong Historical Aircraft Association, https://gwulo.com/atom/13776. Courtesy of the Hong Kong Historical Aircraft Association.

the Chinese flagship magazine of Cathay-MP&GI (Motion Picture & General Investment), he attempted to alleviate the charge of racism made against the film. Chiao told the reporter that he had convinced Gilbert to modify provocative dialogue to avoid racial discrimination against Chinese. Welles originally had a line, "I would rather die than be caught giving out this ferry to you 'Dirty Chinese.'" On Chiao's advice, "Dirty Chinese" was changed to "Dirty Oriental."[84] (Today, both terms are equally racial slurs.)

---

84. Roy Chiao 喬宏, "GangAolundu yanchu suoji" 「港澳輪渡」演出瑣記 [Notes on performing in *Ferry to Hong Kong*], International Screen [國際電影], February 1959, 26–28, http://internationalscreen.net/yingshi-detail.asp?id=2391.

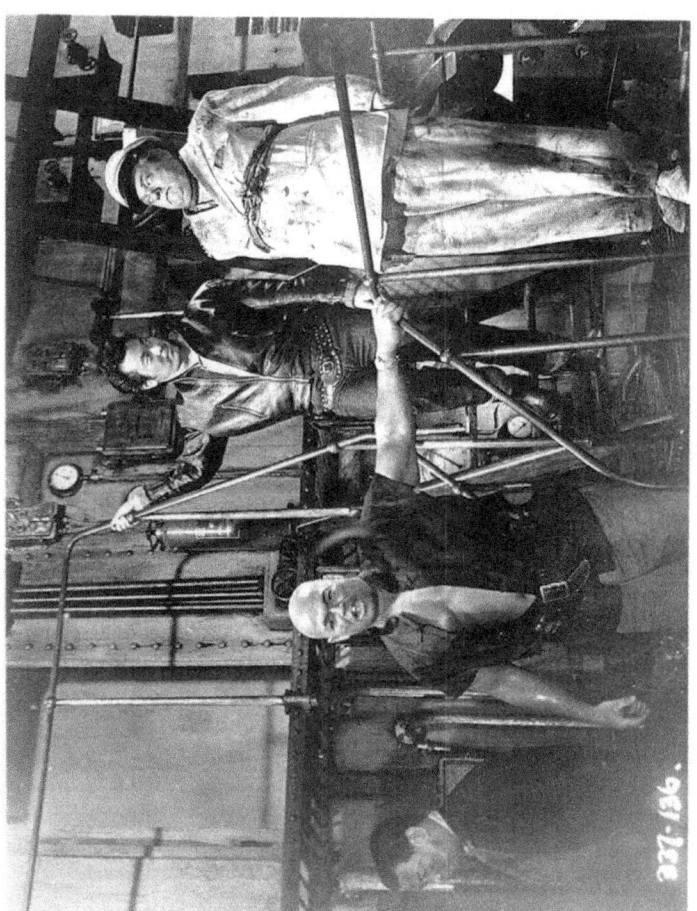

Figure 11: Captain Hart (Orson Welles) captured by Johnny Sing-Up (Roy Chiao) in leather jacket and the bald-headed pirate (British-Indian actor Milton Reid).
Source: Warner Brothers and Seven Arts, Ferry to Hong Kong, printed still. moondollar007 on eBay, https://www.ebay.com/itm/174206463075.

For Marchetti, the root of Hollywood's representation of Asians can be traced back to the "yellow peril" in which these "irresistible, dark, occult forces of the East" will overpower and destroy the Western civilizations, cultures, and values.[85] *Ferry to Hong Kong* smacks of blatant racism comparable to Hollywood's in depicting Chinese as hoodlums and looters. In 1960, it might seem early to be calling out Hollywood's stock representation of Chinese pirates.[86] But Chinese commentators were already furious at the film's Hollywood-style Orientalist gaze. A critic writing for *Chinese Student Weekly* denounced the film's overlooking of Hong Kong's urban modernity as the camera scouted only for exotic scenes.[87] The leftist *Ta Kung Pao* lamented the portrayal of Chinese as violent pirates or more generally as ignorant, unflattering, and primitive subjects. The film inaccurately conveyed an impression of an underdeveloped city with messy traffic and an ineffective police force.[88] Even expatriate viewers in Hong Kong complained about the spectacle of the harbor and the waterfront showcasing backwardness, disorder, and crime: "The only onshore impression given of Hong Kong is a place of drunken brawls in bars full of doubtful girls frequented by sailors of many countries." The sinking of the Fat Annie in Victoria Harbor surrounded by sampans "gives a particular insulting impression of Hong Kong."[89] The film's stereotypical depiction of the pirates as barbaric subjects is aligned with the

---

85. Marchetti, *Romance and the "Yellow Peril,"* 2.

86. *Pirates of the Caribbean: At World's End* (dir. Gore Verbinski, 2007) featured Hong Kong star Chow Yun-fat in the role of a Chinese pirate captain who is bald and scar-faced and who wears a long beard and long nails. Chow's screen time was slashed in half by mainland Chinese censors as defacing the Chinese. See "Disney's 'Pirates 3' Slashed in China," *China Daily*, June 15, 2007, https://www.chinadaily.com.cn/china/2007-06/15/content_895296.htm

87. Danny 但尼, "Yingping: GangAolundu" 影評:「港澳輪渡」 [Film review: *Ferry to Hong Kong*], *Chinese Student Weekly* [中國學生周報], no. 390, January 8, 1960, https://hklit.lib.cuhk.edu.hk/search/?query=%22%E6%B8%AF%E6%BE%B3%E8%BC%AA%E6%B8%A1%22&publish_year_=%5B1920%2C2022%5D.

88. Lin Si 林思, "Yiwushichu: Ping GangAolundu" 一無是處: 評港澳輪渡 [Completely wrong: Review of *Ferry to Hong Kong*], *Ta Kung Pao* [大公報], January 4, 1960.

89. C. M. Faure, "To the Editor: Ferry to Hongkong," *South China Morning Post*, January 4, 1960, 11.

Cold War's "siege mentality" and "Red Scare," in which the Communist enemy is treated as a dark force poised to invade the Free World.[90]

The visual-ethnographic representation of Macau was outrageous to Chinese critics too. The scenic beauty of Macau was inauthentic, and the place where it was filmed was "worse than every postcard in the market."[91] The camera featured a funeral procession. The scene was choreographed with Indigenous Chinese people dressed in archaic Qing dynasty attire to appeal to the Western preconception of premodern China. In real life, Welles took his own camera to film documentaries in Macau, including a Chinese funeral.[92] Adding to the thrill of the actual filming in Macau was a potential confrontation of the crew with angry Communist supporters when they did location shooting close to the Chinese (Zhuhai) border. Communist agitators across the shore warned the crew to keep off Chinese territory, waving red flags and playing propaganda songs at Gilbert's teammates. In response, the director broadcasted a humorous speech and played jazz music. Gilbert yelled at the Communist activists across the shore: "If you have to arrest someone, go for my cinematographer!"[93]

In the same year when *Ferry to Hong Kong* began its world premiere in London and Hong Kong, the MP&GI-produced Mandarin picture *Air Hostess* (1959) was released in Hong Kong, starring Chiao and Grace Chang, a talented Hong Kong Chinese movie star, singer, and popular idol. MP&GI followed the Hollywood studio system and vertical integration of production, nurturing its movie stars and churning out romantic

90. Paul B. Rich, *Cinema and Unconventional Warfare in the Twentieth Century Insurgency, Terrorism and Special Operations* (London: Bloomsbury Academic, 2018), 31–35.

91. Lin, "Yiwushichu: Ping GangAolundu."

92. Lewin, "Third Man in Hong Kong Part 1."

93. "Gongfei gejie yaoqi daoluan GangAolundu paishe gongzuo" 共匪隔界搖旗搗亂 港澳輪渡拍攝工作 [Commies wave flags and mess with *Ferry to Hong Kong* filming over the border], *Central Daily News* [中央日報], January 8, 1959; Hung 虹, "'GangAolundu' de daoyan Jierbo touzi zhipian"「港澳輪渡」的導演 基爾勃投資製片 [Director of *Ferry to Hong Kong*: Gilbert invests in own production], *United Daily News* [聯合報], February 24, 1961.

comedies and urban dramas often filled with North American pop music and dance numbers such as jazz, cha-cha, and mambo. A big-budget film shot in Eastman Color, *Air Hostess* was shot like a travelogue, showcasing the exotic beauty and modernity of the Asian cities of Hong Kong, Singapore, Bangkok, and Taipei. The cinematic tourism of Asian and Southeast Asian locations outside the People's Republic of China (PRC) orbit created an inter-Asian, pan-Chinese imagined community free from the reach of Communism and celebrated the US-supported modernity of Taiwan as an exemplar of Free China. Poshek Fu claims that *Air Hostess* was one of the most significant Hong Kong Cold War movies.[94] The high-cost, high-tech, star-studded transnational production served ideologically to highlight the desire of Asian regions for technomodernity (air travel and mobility) and a good life of capitalist prosperity for pan-Chinese societies outside of mainland China.

While MP&GI exploited the screen success and widespread idolization of its stars to promote the tourist gaze of Asian modernity for pan-Chinese audiences, the Asian actors were marginalized in Hollywood and Anglo-American cinema. Chang was relegated to a minor, uncredited role as a sampan girl in *Soldier of Fortune*. *Ferry to Hong Kong* denigrated Chiao by casting him as a small-time rogue. It did not attend to ethnic authenticity of its Chinese characters or care very much about taking a share of the Asian market. The film was a journey of British Orientalism and cinematic tourism in search of a lost prestige, as well as an exercise in self-portraiture when faced with the oriental Other.

Besides MP&GI's coverage of Chiao's role in *Ferry to Hong Kong*, Taiwan's *Central Daily News* and *United Daily News* applauded the Hong Kong actor for working with international stars; it brought prestige, regardless of the fact that he played an Asian villain. He was called "the actor of our

---

94. Poshek Fu, "Entertainment and Propaganda: Hong Kong Cinema and Asia's Cold War," in *The Cold War and Asian Cinemas*, eds. Poshek Fu and Man-Fung Yip (New York: Routledge, 2019), 254–57.

country (China)" (我國演員) and the "actor of China" (中國演員).[95] The claim to "China" by the Taiwan presses unabashedly unveiled the feud between the Guomindang (GMD) regime in Taiwan and the PRC over being the legitimate representative of China in the 1960s. In Hong Kong, the British authorities were cautious to avoid any possibility of a "Two Chinas" situation and strove to prevent the Taiwan question from triggering a hot war in East Asia.[96] In the local cinema, censors were instructed to closely examine all Chinese-language pictures produced in mainland China, Taiwan, and Hong Kong and to "bear in mind particular sensitivities on both sides of the camp to implied recognition of 'Two Chinas.' "[97]

The Taiwanese press was spinning the story to suit its own political agenda. Its claim of legitimacy of "China" indicated how Cold War geopolitics played out in transnational production. Indeed, the corporate strategies of MP&GI meshed with Cold War cultural politics. MP&GI's Malayan Chinese owner, Dato Loke Wan Tho, was involved in the politics of anti-Communism and "identified with 'Free China'–Taiwan—as the custodian of Chinese culture and thereby the only legitimate Chinese government."[98] When anti-Communist messages were taboo under Hong Kong's censorship, MP&GI turned to the Hollywood model to produce

---

95. Hao 豪, "Aoxunweiersi yu Kouyouningsi hezuo GangAolundu" 奧遜威爾斯與寇尤寧斯 合作「港澳輪渡」 [Orson Welles and Curt Jurgens collaborate in *Ferry to Hong Kong*], *Central Daily News* [中央日報], January 9, 1961; " 'GangAolundu' paishe wancheng Kouyouningsi xie furen budu miyue Aoxunweiersi zuo sanlunche youjie" 「港澳輪渡」拍攝完成 寇尤甯斯偕夫人補渡蜜月　奧遜威爾斯坐三輪車遊街 [*Ferry to Hong Kong* filming completed: Curt Jurgens' couple makes up honeymoon; Orson Welles sightsees on trishaw], *United Daily News* [聯合報], February 25, 1959; Hung, " 'Gangaolundu' de daoyan Jierbo touzi zhipian."

96. Steve Yui-Sang Tsang, "Unwitting Partners: Relations between Taiwan and Britain, 1950–1958," *East Asian History*, no. 7 (June 1994): 105–6.

97. "A Statement of the General Principles as adopted on 20 November 1965 by the Film Censorship Board of Review," Hong Kong record series, Hong Kong Public Records Office, HKRS 934–5–34, 1965.

98. Poshek Fu, "Modernity, Diasporic Capital, and 1950's Hong Kong Mandarin Cinema," *Jump Cut: A Review of Contemporary Cinema*, no. 49 (Spring 2007), https://www.ejump cut.org/archive/jc49.2007/Poshek/.

urbanized romances and narratives of Free Asia modernity. Hollywood has become a channel for exporting American capitalism, democratic freedom, and modern-life culture. MP&GI then promoted one of its biggest movie stars, Grace Chang, to American television.[99] Yet, this regional Asian cinematic pursuit of modernity was juxtaposed with the tradition of Orientalism. As mentioned, behind the filming and screening event of *A Ferry to Hong Kong*, the involvement of Chinese/Asian stars in Anglo-American big-budget productions concealed the Cold War geopolitics and the lopsided power dynamics of Hollywood and Asian cinema.

Coda: Sinking Boat and Floating City

Ferry to Hong Kong only played in Hong Kong's cinemas for ten days, starting on the last day of 1959.[100] It failed to sell the British story to American audiences. The picture was belatedly screened across the United States in 1961 and was considered "so terrible that managers refused first-run booking."[101] One US exhibitor called the film a complete disaster: "Two real good actors but couldn't get any business. Played one day to a loss."[102] Even British scholars have considered the film to be "totally unbelievable." Unsurprisingly, "it was quickly shunted into oblivion."[103] Costing half a million pounds (roughly $1.4 million), and with an all-star extravaganza, the

---

99. For Grace Chang's appearance on an American television show that conformed to the exoticized view of Chinese womanhood, see Stacilee Ford, "'Reel Sisters' and Other Diplomacy: Cathay Studios and Cold War Cultural Production," in *Hong Kong in the Cold War*, 183–88.

100. Adam Nebbs, "*Ferry to Hong Kong*: The 1959 Film about a Stateless Passenger Who Spent 10 Months Aboard the *Lee Hong*," *South China Morning Post*, May 15, 2020, https://www.scmp.com/magazines/post-magazine/travel/article/3084137/ferry-hong-kong-1959-film-about-stateless-passenger.

101. Richard L. Coe, "The Circuit Riders," *Washington Post*, October 29, 1961, sec. G.

102. Mel Danner, "The Exhibitor Has His Say about Pictures," Boxoffice, December 18, 1961, https://lantern.mediahist.org/catalog/boxofficeoctdec180boxo_0616.

103. Harper and Porter, *British Cinema in the 1950s*, 55.

international production split at the seams as the British and American part-ners quarreled.

The film was a dire flop across the globe not only because of the botched Anglo-American cooperation but also because the demands of entertain-ment and propaganda clashed. They fell out of place in the film's storytell-ing. How can a film inform and instruct an audience given the yearning for romance and spectacle, and above all for a respite from politics, in mass entertainments? The romantic and melodramatic plots of the British film proved fruitless in appealing to an international audience; it did not do par-ticularly well even at home, receiving terrible reviews. Gilbert would later direct three James Bond movies: *You Only Live Twice* (1967), *The Spy Who Loved Me* (1977), and *Moonraker* (1979). They were internationally more popular and successful as Cold War espionage cinema. *Ferry to Hong Kong* might well have failed to become a popular success at home and abroad despite a stellar cast and spectacular promotions. The film could be seen as a British cultural-diplomatic response to reestablish through cinematic soft power national assurance on the Cold War international stage.

Situated at the end of the 1950s, *Ferry to Hong Kong* invites a symp-tomatic reading. It tells how Conrad saves the day, overpowering the Com-munist intruders, and jumps to rescue the sinking vessel teeming with helpless Chinese passengers. Betraying the traits of arrogance, callousness, and snobbery, Captain Hart is a stand-in for the pragmatic colonial men-tality, as he hesitates to save the people on the stranded junks.[104] Conrad, the ordinary man with a bruised soul, reacts without hesitation to the most fateful conditions. He supersedes the morally weak captain to call for a rescue of the drifting survivors. The downtrodden loner is eventually recast as the "colonial freedom fighter"—with the charisma of "audacity, stoicism,

---

104. I submit to Hazel Shu Chen's view and further explain that the role of Welles as a stub-born British captain, administrator, and buffoon in the film must have disappointed Hollywood's audience then and now, for whom Welles on-screen had already embodied the identity of Americans who should be saving the world from Communism and co-lonialism.

unflappability, good manners, a sense of duty, and an ability to command" to defeat the Communist intruders.[105]

Ferry to Hong Kong is more than an adventurous picture of a vagabond who tries to redeem himself and salvage self-respect and a future from the wreck as the ferry goes down into the sea. At the end, Conrad chooses to depart from Ferrers after their romantic adventure, but he promises her school children that he is going to "fight the dragon" before coming back to "claim the princess." Conrad uses the medieval tale of the chivalric knight slaying the "dragon" to avow his moral mission to continue the fight against Dragon China. The romantic narrative is sacrificed to make way for the film's political intent. The antihero becomes the last British white knight of the fallen empire—with its lost transnational dominance and international supremacy—who struggles to regenerate the British position as the guardian of the Free World with heroism, nostalgia, and anxiety. But despite the denigration of imperialist ambition on the part of the protagonist, the film indulges in an inconspicuous form of imperial nostalgia by evoking the thrill of a battle between good and evil, the glamour of the gentleman-hero, and exotic far-flung locations

The analogy of the sinking ferry as a broken empire may not be easily traced in this transnational production packaged as a suspense trip and romance. However, the screen adaptation makes a pivotal change to the novel, in which the *Fat Annie* survived the raids and storms. As the protagonist Clarry in the novel ruminates on the battered ferryboat:

> He [Clarry] stood on the deck and took a long, faintly startled look at the Fat Annie, wondering for the first time—now that it was safe to wonder—how she had survived. She glimmered greyly, crusted with salt all the way down from the dented smoke-stack to the saloon. She looked as though

---

105. Stuart Ward, "Introduction," in *British Culture and the End of Empire*, 15. For a discussion of British imperial heroes depicted in the post-imperial films, see Jeffrey Richards, "Imperial Heroes for a Post-imperial Age: Films and the End of Empire," in *British Culture and the End of Empire*, 128–44.

she had just emerged exhaustedly from a naval battle, with bent stanchions, battered bulwarks, an appalling scum of debris covering her deck.[106]

The military and political undertones could not be more obvious in this passage. The *Fat Annie* is not crushed by the (Communist Chinese) enemy: "She'll never sink, the devil won't have her!" On the screen, Captain Hart and Conrad on the lifeboat witness the vessel dipping slowly into the sea in a wide-angle shot. Just a moment before, Hart still wanted to keep his ship afloat: "Come on, Annie. Come on, old lady," Hart declares. "This is still my ship!" We are aware of the captain's dignified and tragic tone and understand why Welles's somewhat comic performance could have spoiled the film throughout. The ending dwells on the moment of the sinking vessel and the image of abject defeat. "She's lived quite a whole of our life," Hart moans, noting their shared fate with the sunken ferry: "Eternal trying—you [Conrad], me, and the *Fat Annie*." The loss of the anthropomorphized ship is endowed with connotations of self-sacrifice and the decline of the empire.

The emblem of the sinking ship is intrinsic to the postwar naval films of Britain, an island nation and a bygone maritime empire. Jonathan Rayner identifies the status of the warship in naval films as the symbol of the nation and explains how the depiction of its loss appears dangerously defeatist to British self-esteem.[107] The sinking boat imagery serves as a traumatic device to recall British naval war stories and retell romantic and narcissistic tales of British gallantry and international influence in global politics. The scene of Conrad saving the Chinese refugees on abandoned junks to board the ferry could easily remind British viewers of *Dunkirk*, a film about the historical evacuation of allied soldiers from northern France in World War II. The two British films were made just one year apart.[108] Dunkirk, a gigantic retreat if

---

106. Kent, *Ferry to Hongkong*, 267.
107. Jonathan Rayner, *The Naval War Film: Genre, History and National Cinema* (Manchester: Manchester University Press, 2007), 215.
108. The remake of *Dunkirk* (2017) by British American director Christopher Nolan indicates that nostalgic war films have continued to thrive at the British box office.

not defeat, has become a wartime icon and national event through cinema-tization, in which the sea is central. This would account for Rank's choice of the novel to narrate the British Hong Kong journey as a watered-down naval story instead of a story on land.

Whereas the in-between territory of the China Sea connotes the underground routes of pirates, smugglers, and refugees, the drifting *Fat Annie* is overloaded with significations of the Cold War city. The mass exodus of Chinese refugees fleeing Communist China to colonial Hong Kong in the early 1950s brought serious social and economic problems to the city as well as a strong propaganda advantage for the Free World.[109] It forced the colonial government to impose border controls to stem the huge flow of illegal migrants. Cinematically, the floating ferry at sea is reflective of the class and power structure of Hong Kong society (West-erners and rich Chinese on the upper deck; Chinese commoners, coo-lies, and a disgraced Conrad on the lower deck) and emblematic of the precarity of Hong Kong as the floating city embattled in its Cold War dilemmas. In a nostalgic and eulogistic mode, the film projects images of benign British colonials. Ferrers, the schoolteacher, behaves like a missionary, taking care of her well-educated diasporic Chinese students. Conrad shows his respect for the coffin-carrying fellow passengers, aghast at the cruelty of the (Communist) pirates who throw the coffin, with the body inside, into the sea.

A neglected cinematic fragment, *Ferry to Hong Kong* offers a sound tes-timony of British sentiments and ideology vis-a-vis the geopolitics between Hong Kong, Britain, and Communist China during the Cold War. Postwar Hong Kong was becoming prosperous as well as precarious. A Chinese take-over of the city, as colonial officials and merchants had realized much earlier in the 1950s, seemed distant but was doomed to happen: "They feel that

---

109. Glen Peterson, "To Be or Not to Be a Refugee: The International Politic of the Hong Kong Refugee Crisis, 1949–55," *Journal of Imperial and Commonwealth History* 36, no. 2 (2008): 171–95.

their island is on Red China's time-table."[110] Fast forward to the late 1980s, in the run-up to Hong Kong's transfer of sovereignty in 1997, the moral question remained for British intellectuals: "Should the British government hand over a city of six million people, among whom a large number are British subjects, to a regime not known for its tolerance of individual liberty and political dissent?"[111]

History may not repeat itself, but sometimes it rhymes.[112] The former colonial regime has now been receiving up to one hundred thousand Hong Kong residents who sought to escape from Communist rule under the "one country, two systems" policy.[113] The modern-day Conrad resumes his moral responsibility with pragmatism, humanitarianism, and imperial nostalgia. The untold story of *A Ferry to Hong Kong* still resonates with our current anxiety and reiterates the unknown nature of the future. It is—for sure—to be continued.

Acknowledgments

This article was written with the support of the Research Grants Council of Hong Kong. I am deeply grateful to Lee Sing Chak, Kuan Chee Wah, and Jessica Siu-yin Yeung for their research assistance. I am indebted to Isabel Galwey, Hazel Shu Chen, Mike Ingham, and my anonymous reviewers for their comments.

110. John G Norris, "Hong Kong Low on Reds' 'List,'" *Washington Post and Times Herald*, December 5, 1958.

111. James T. H. Tang, "From Empire Defence to Imperial Retreat: Britain's Postwar China Policy and the Decolonization of Hong Kong," *Modern Asian Studies* 28, no. 2 (1994): 337.

112. "History does not repeat itself, but it rhymes"—the humorous aphorism is usually attributed to Mark Twain, but this ascription remains dubious.

113. Laura Westbrook and Danny Mok, "Britain Plans Extension of BN(O) Visa Scheme to Allow Hongkongers Aged 18 to 24 to Apply Independently of Parents," *South China Morning Post*, February 24, 2022, https://www.scmp.com/news/hong-kong/society/article/3168295/emigration-wave-continues-about-100000-hongkongers-apply-bno.

Imagining Cooperation

Cold War Aesthetics for a Hot Planet

MARINA KANETI

Abstract

What does cooperation between rival superpowers look like? Do global issues have the capacity to rise above the geopolitics of the day and trigger alignment between rival powers? This paper argues the Cold War joint space exploration program between the United States and USSR provides a lesson on the limits of cooperation. These limits, I suggest, are not only a matter of power preferences, institutional differences, material disincentives, or even a consequence of a tendency for free-riding. Rather, they are also the result of incompatible "common sense" perceptions. Cooperation, even if institutionally viable, as in the case of the joint space program, can be constrained due to a lack of popular endorsement and legitimacy.

To develop the argument, I examine the aesthetics of cooperation rendered through widely circulated media images associated with space cooperation during the Cold War. I argue the Cold War imaginary can serve as both critique and inspiration for today's attempts to legitimize cooperation on global issues such as climate change. It provides insights on the role of "common sense" perceptions and the ways in which they inform questions concerning universality, the role of affect, and the alure of competition.

Keywords: Cooperation, Cold War, Visual politics, global issues, legitimacy

With US-China relations at an all-time low, and a US administration persistently keen on drawing a line between democracies and autocracies, cooperation on issues of global concern, such as the climate crisis, has become

https://doi.org/10.3998/gs.2512

a nonentity, falling off completely from public view. Yet, at a time of re-cord-breaking temperatures, melting ice caps, and warnings of a "ghastly future of mass extinction, declining health and climate-disruption upheav-als,"[1] shouldn't an issue with global ramifications be a top priority for the two leading superpowers? How is their noncooperation legitimized? Or, is cooperation simply unimaginable?

In the standard oeuvre of academic literature, political science, and international relations in particular, the debate around global coopera-tion often centers on various aspects of rationality and power. Realism posits that cooperation between states, if at all possible, is a reflection of power distribution and contingent on (rational) strategic preferenc-es.[2] Noncooperation, in turn, is the result of strategic choices made in pursuit of a preferred national agenda; a refusal to cooperate does not have to be justified beyond the fact that it might contradict such strategic preferences. In fact, and according to the realists, entirely the opposite is true: opposition, and the capacity to compete, become the ultimate signs of power and capability. For proponents of liberalism, while power and strategic preferences matter, cooperation is seen as the product of institutional arrangements.[3] To this end, a failure to cooperate can be legitimated on the grounds of inadequate institutional mechanisms. For constructivists, cooperation is possible on the basis of values, ideas, and shared norms.[4] Noncooperation would therefore be the natural result of different values and ideas: the aforementioned distinction between democracies and autocracies can result in cooperation failure undergird-ing ontological differences on a matter of great significance to the two

1. P. Weston, "Top Scientists Warn of 'Ghastly Future of Mass Extinction' and Climate Dis-ruption," *Guardian*, January 13, 2021, https://www.theguardian.com/environment/2021/jan/13/top-scientists-warn-of-ghastly-future-of-mass-extinction-and-climate-disruption-aoe.
2. See especially K. N. Waltz, *Realism and International Politics* (New York: Routledge, 2008).
3. R. O. Keohane and J. S. Nye, *Power and Interdependence* (Boston: Longman, 2012).
4. A. Wendt, *Social Theory of International Politics* (Cambridge: Cambridge University Press, 1999).

opposing sides. Beyond these standard accounts, a number of scholars have also conceptualized the question of cooperation around global concerns from the vantage point of the issue, or object, itself. Namely, for proponents of object-centered theories, cooperation is possible because of the universal nature of the issue itself.[5] Accordingly, the need for cooperation emerges not because of actors' individual preferences, rationalities, or power positions but because issues with potentially universal impact (such as climate change or poverty) create a space for cooperation and channel a collective action approach beyond individual government preferences.[6] Such accounts, while insightful, assume societal perceptions converge with government positions: be those in pursuit of allegedly universal challenges or for the purposes of reasserting state power. As such, these accounts tend to overlook questions of legitimacy and "common sense" social perceptions.

This article suggests the need for a new approach to cooperation, one which considers how broad-based "common sense" perceptions and affective reactions provide orientation and legitimacy to collaborative action. Certainly, a choice of cooperation or competition in the global arena might be entirely of a government's choosing. Yet, this article argues such choices still demand legitimacy and social acceptance. Cooperation, especially at times of extreme ideological confrontation, needs to align with an overarching common sense understanding of threats and opportunities. Indeed, the need to legitimize government actions is equally valid in both democratic and non-democratic settings. For example, as Elizabeth Perry has

---

5. H. Bulkeley, *Accomplishing Climate Governance* (Cambridge: Cambridge University Press, 2016); O. Corry, *Constructing a Global Polity: Theory, Discourse and Governance* (New York: Palgrave Macmillan, 2013); B. B. Allan, "Producing the Climate: States, Scientists, and the Constitution of Global Governance Objects," *International organization* 71 (2017): 131–62.

6. See, for example, the work of Clark Miller for an argument on how the constitution of the climate helped to produce the very idea of global governance: C. A. Miller, *Climate Science and the Making of a Global Political Order* (London and New York: Routledge, 2004).

masterfully argued, the Chinese government resorts to a sophisticated form of cultural governance in order to legitimize its political authority.[7] This is enabled not only by perpetually asserting historical and cultural narratives and images in the public sphere but also by engaging in the meticulous work of gauging and guiding public sentiment. As such, the question of legitimacy is a question of visual politics whereby the realm of visuality and the common sense perceptions it informs, play a critical role in the overarching viability and perpetuity of the ruling regime.

To develop the argument, the article proposes a phenomenological reading on the aesthetics of cooperation. I briefly elaborate on the choice of both aesthetics and phenomenology. Building on Jacques Rancière's conceptualization of the politics of aesthetics,[8] my examination concerns the visual politics of cooperation: how cooperation is seen through various representations and can therefore be felt, thought of, and accepted as common sense understanding. The proposition that visuality and images play a major role in international relations and inform every aspect of social interactions, emotions, and thinking is not new. Images, as W. J. T Mitchell says, are "active players in the game of establishing and changing values."[9] Over the past two decades, a growing engagement with visuality and the politics of aesthetics has enriched and complicated the study of international relations. Here, *aesthetics* is not meant to signify a study of beauty. Instead, it concerns the realm of visibility: what can be seen and can therefore be sensed, felt, thought, and accepted as "common sense." An interrogation of the politics of aesthetics, therefore, entails a critical exploration of how visuality sculpts our collective values, our perceptions, and understandings of what is permissible, legible, and "common sensical." Starting from the premise that our knowledge of the world is based on what is made visible and what remains hidden, a critical

7. E. J. Perry, "Cultural Governance in Contemporary China: 'Re-orienting' Party Propaganda," in V. Shue and P. Thoronton, *To Govern China: Evolving Practices of Power* (Cambridge: Cambridge University Press, 2017).
8. J. Rancière, *The Politics of Aesthetics* (London: Bloomsbury Academic, 2013).
9. J. T. Mitchell, ed., *Pictorial Turn* (London: Taylor and Francis, 2018).

engagement with the politics of aesthetics and the role of images allows for interrogation of our ways of looking, seeing, evaluating, and thinking frame questions of legitimacy. It allows for exploration of how the realm of visuals delineates what is "sensible" and, by extension, what is thinkable, meaningful, valuable, and acceptable. In other words, looking at interactions from the politics of aesthetics prespective enables interrogation of the power of images in constituting collective notions of "common sense" and what are the "conditions of possibilities" for potential transformation of values, affect, and thoughts by making visible and sayable alternative realities.[10]

Using such insights, a number of international relations scholars have argued that visual artifacts not only depict politics,[11] but can also shape collective understanding of issues ranging from violence and security to sovereignty and trauma.[12] One such example comes from the massive public outcry against the US War on Terror. This outcry was triggered as a response to the widespread circulation of graphic photographs of torture by US troops of detainees at the Abu Ghraib prison facility. Prior to the graphic images, there was ample knowledge of both the War on Terror and torture techniques but no public attention given to the issue. It was the horrific visual displays of abuse and maltreatment that sparked international outrage and demands for accountability, including US Congressional hearings and investigations.[13] Even if there was no significant alteration of US foreign policies as a result; globally, the images came to be recognized as symbols of US abuse of power and loss of legitimacy.

10. R. Bleiker, *Visual Global Politics* (London: Taylor and Francis, 2018).
11. R. Bleiker, "Writing Visual Global Politics: in Defence of a Pluralist Approach—A Response to Gabi Schlag, 'Thinking and Writing Visual Global Politics,'" *International Journal of Politics, Culture, and Society* 32 (2019): 115–23.
12. See, for example, J. Vuori and R. Saugmann, *Visual Security Studies: Sights and Spectacles of Insecurity and War* Routledge Taylor and Francis, 2018; R. Adler-Nissen, K. E. Andersen, and L. Hansen, "Images, Emotions, and International Politics: The Death of Alan Kurdi," *Review of International Studies* 46 (2020): 75–95; K. Grayson and J. Mawdsley, "Scopic Regimes and the Visual Turn in International Relations: Seeing World Politics through the Drone," *European Journal of International Relations* 25 (2020): 431–57.
13. US Cong., Resolution of Inquiry Regarding Pictures, H. Rept. No. 108–547 (2004).

In turning to images, I also combine the politics of aesthetics perspective with the concept of (dis)orientation, developed most prominently in Sara Ahmed's *Queer Phenomenology*.[14] As Ahmed argues, with phenomenology we can interrogate how norms or orientations "shape not only how we inhabit space, but how we apprehend this world of shared inhabitance, as well as 'who' or 'what' we direct our energy and attention toward."[15] Ahmed's interrogation of how we come to be oriented; how orientations come to be; how they are revealed, obscured, and interpreted speak directly to the complex ways in which common sense perceptions come into being. Ahmed's insights therefore allow me to consider the connections between common sense and orientation, and how visuality and sight become the source of (dis)orientation, affecting notions of identity, reality, and universal values. Approaching international cooperation from the vantage point of images and the politics of aesthetics, the article explores the sights and sites that provide orientation, sculpt different common sense understandings, and contribute to notions of legitimacy. A turn to visuality, and an interrogation of how visuality can "turn" us, alerts us to the ways in which cooperation is as much a product of affective, cultural, and political realities as it is the result of strategic interests, global dominance agendas, universal values, and technological aspirations.[16]

Analyzing moments of (dis)orientation and the construction of common sense perceptions, this article will first elaborate on the argument by exploring the imaginary of cooperation and universal values in a setting of extreme hostility and ideological confrontation: the US-USSR space exploration during the Cold War. The article will then briefly explore the ongoing tussle between the United States and China and consider some Cold War insights in relation to the current prospects for cooperation on the climate crisis.

---

14. S. Ahmed, *Queer Phenomenology: Orientations, Objects, Others* (Durham, NC: Duke University Press, 2006).
15. Ahmed, *Queer Phenomenology*, 3.
16. See also H. McCurdy, *Space and the American Imagination*, 2nd ed. (Baltimore: Johns Hopkins University Press, 2011).

Certainly, the Cold War and the present moment are not fully comparable in terms of scope, timing, and nature of interactions, and neither are contemporary challenges akin to the dynamics of bipolarity more than half a century ago. Yet, turning to the Cold War period provides a fruitful way of exploring not only the possibilities for cooperation on an issue of global proportions but also whether or how such cooperation was able to find legitimacy: how it was imagined and made "common sensical" at a time of extreme hostility and distrust toward the opposite side. The article proceeds in three parts. It first explores the aesthetics of Cold War space cooperation as represented in two popular magazines—the American *Time* magazine[17] and the Soviet *Krokodil*.[18] Next, the article maps out the aesthetics of cooperation emerging from Cold War imagery onto contemporary representations of climate cooperation between the United States and China. The final section draws conclusions on the possibilities for cooperation and the insights phenomenology can contribute to the study of legitimacy and societal common sense orientations.

Dining with the Enemy

Contrary to popular renderings of the past, the period of the Cold War did indeed mark the start of a prominent cooperation between the United States and the USSR on an issue of global proportions and significance: space exploration. Despite considerable challenges over the decades, collaborative space exploration outlived ideological hostilities between the rival governments and continued even beyond the existence of the USSR. The space program seemingly reaffirmed the notion that global issues have the capacity

---

17. *Time* Magazine was founded in 1923. The distinctive cover featuring a prominent image associated with current news events was introduced in 1927.
18. *Krokodil* was founded in 1922 and was published once a week. *Krokodil* used caricature and visuals to lampoon political figures and events. It discontinued publication after the collapse of the Soviet Union (apart from a brief reinstatement from 2005 to 2008).

to rise above the geopolitics of the day. Accordingly, what allegedly made co-operation in space legitimate was the universal nature of space itself and the fact that its unknown vastness could not be tackled by one country alone. At the same time, collaboration would ensure that this spatial universe benefits the entirety of humanity, not individual nations.[19] Yet, the US-USSR space exploration program also remains the source of a curious paradox: although the cooperation itself began as early as the 1960s and continued for many decades after the collapse of the Soviet Union, the popular imaginary and memories of the period tend to focus on key moments of space competition—nearly erasing any collective memory of cooperation. Indeed, along with the very notion of the Cold War itself, the two countries' space interactions are remembered from the prism of competition rather than co-operation. Consider how, in the prevailing common sense imaginary and popular memory the US-USSR space interactions are typically associated with "the Space Race" and "Star Wars" and relegated to tropes such as the "First Man on the Moon," "First Man in Orbit," "Trailblazers/Pioneers in Space," etc. Similarly, popular magazines, cartoons, and posters consistently render space as yet another stage for great power competition, a grand contest in scientific and technological prowess and superiority.[20] Today, online search engines also produce many more results and images for "Cold War space competition" than for "Cold War space cooperation." And there is also the naming: "The Space Race" features prominently on the NASA website and various study materials of the time. In comparison, the singular reference to

---

19. This was also Joe Biden's statement after a bilateral summit meeting with Vladimir Putin in 2021; also, see M. Luxmoore, "U.S. and Russia Find Some Common Ground in Space," Foreign Policy, November 3, 2021, https://foreignpolicy.com/2021/11/03/us-russia-space-cooperation-nasa-sirius/.

20. One recent example of this was the 2016 exhibition in Moscow entitled *Вперёд! К звёздам!* (Advance! Towards the stars) featuring posters from the Soviet space program from the 1950s on. Among the forty posters, there is not one image of the space cooperation with the United States. See "Exhibition: 'Forward! To the Stars!' Museum of Political History of Russia, accessed October 26, 2022, http://collectiononline.polithistory.ru/entity/EXHIBITION/3942744 Accessed March 2022.

cooperation, known as "The Handshake in Space," appears as a momentary, short-lived event, hardly of the same magnitude and significance as the decades-long space competition.

I suggest that this collective amnesia of the factual existence of space cooperation during the Cold War is a matter of affective orientation, driven by both the desire and ability to accept the possibility for cooperation, to see it as a legitimate option. Cooperation is contingent on affect, which, in turn, forms the core nucleus of cooperation: trust and assurance.[21] Yet, it is precisely trust and assurance that tend to be unimaginable at a time of extreme ideological differences. Hence also the condition of affective disorientation, of blocking, and refusing to acknowledge the significance of factual evidence. This is also how, even if cooperation exists on an institutional level, the common sense perception is still framed by representations that both explicitly and implicitly carry a message of distrust and suspicion. As such, competition, rather than cooperation, remains the visible aesthetic frame, (re)orienting perceptions on the range of interactions between opposing powers. This type of (re)orientation is visible not only in the abundant proliferation of images of space competition but even in images allegedly associated with cooperation.

Consider, for example, the representation of a key moment in US-USSR space cooperation: the 1975 Apollo-Soyuz mission. Most commonly known as the "Handshake in Space," both the mission and associated collaboration were meant to be a grand spectacle. An unprecedented feat of technological mastery, the joint enterprise marked the first time when astronauts from different spacecraft could physically meet and interact outside the Earth. Betting on a grand spectacle, both the United States and the USSR streamed the event—the entire process down to the joint dinner in space—live on television. Meticulous details around the staging of what *Time* magazine named a "Space Spectacular" included a lengthy negotiation involving considerations such as the exact location over the Earth's surface where the two

---

21. E. Ostrom, *Governing the commons: the evolution of institutions for collective action* (Cambridge: Cambridge University Press, 2015).

spacecraft would intersect and activate the docking platform, details on the meals the two crews would share, the body positioning so that the TV cameras could capture the actual handshake between the two sides, etc.

By all accounts, the Apollo-Soyuz mission marked a moment of unprecedented political achievement, more than a technological breakthrough. Certainly, the development of a docking platform to which both ships could latch onto so that both crews could traverse between the two spacecrafts was important. But, according to many who were part of the process, such platform construction did not require the type of advanced technological know-how and knowledge-sharing that more sophisticated, joint space exploration initiatives might entail. Furthermore, the construction of a docking module at the exorbitant cost of US $100 million (or the equivalent of $765 million in 2022) allegedly had limited use after this staged event. In the United States, the revelation of such facts made the entire enterprise quite questionable. Nevertheless, the political significance and effects of the engagement were many. To those involved directly in the launch, including the two crews, the entire process of interaction revealed a host of unexpected similarities.[22] The preparation for the mission humanized a relationship that was otherwise fraught with misinformation and propaganda on both sides. Moreover, the mission greatly benefited from the two sides' sense of "space comradery"[23] and inherent understanding of comparable technological capabilities: each side had had its successes and failures in space exploration, which were equally valuable and informative in crafting the joint mission.

In the aftermath of the joint mission success, both the White House and the Kremlin sought to capitalize on the event as a political opportunity for further cooperation. After hosting the two USSR cosmonauts at the White House, president Gerald Ford argued: "The broader we can make our

---

22. Y. Karash, *The Superpower Odyssey: A Russian Perspective on Space Cooperation* (Reston, VA: American Institute of Aeronautics and Astronautics, 1999).

23. T. Ellis, " 'Howdy Partner!' Space Brotherhood, Detente and the Symbolism of the 1975 Apollo–Soyuz Test Project," *Journal of American Studies* 53 (2019): 744–69.

relations in health, in environment, in space . . . the better it is for us here in America, and for the Soviet Union."[24] In Ford's assessment, the mission opened a door to cooperation not as a singular affair but an exemplar of how politicians should aspire to reach an agreement back on Earth. In his words, "Our astronauts can fit together in the most intricate scientific equipment, work together, and shake hands 137 miles out in space, we as statesmen have an obligation to do as well on Earth."[25]

Accordingly, popular images in both countries followed suit. Aligned with the mood in the White House administration, the cover of the popular Soviet magazine *Krokodil* pictured a "Cold War" caricature squeezed between the two docking spacecraft, helplessly dropping its sword (fig. 1).[26]

One interpretation of the image could be that the entire purpose of the two spacecraft in space was to show how the Cold War was meant to end. Indeed, the sheer force of the two spacecraft joining above the Earth creates the impression that the Cold War had run out of options to rule over the globe and had just lost its weapon/sword of destruction.

Yet, the *Krokodil* image also allows for an alternative, noncelebratory interpretation. In a different way of seeing, the central subject is not space exploration or collaboration. Rather, it is a spectacle of sheer violence. To someone unaware of the joint mission and the attempt to connect the two spacecraft, the image could appear as a moment of collision between two rockets set against one another. This impression is enhanced by the prominent lines depicting the trajectory of movement of the two spacecraft, suggesting not a moment of rest, cojoining, and docking but of an impending crash. The fantastical image of the Cold War stricken by the collision could also be reinterpreted as an attempt by the Cold War "himself" to avoid further escalation of hostilities: if the two spacecraft were to clash, then

24. Quoted in Ellis, "'Howdy Partner!'"
25. J. Naughton, "Ford Bids Nations Live Up to Spirit of Helsinki Pact," *New York Times*, August 2, 1975, 1.
26. "СОЮЗ-АПОЛЛОН. Маневры на орбите" [Soyuz-Apollo: On-orbit maneuvers], Live Journal, accessed October 26, 2022, https://1500py470.livejournal.com/136067.html.

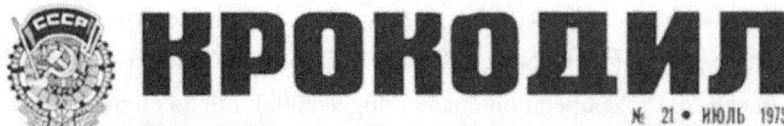

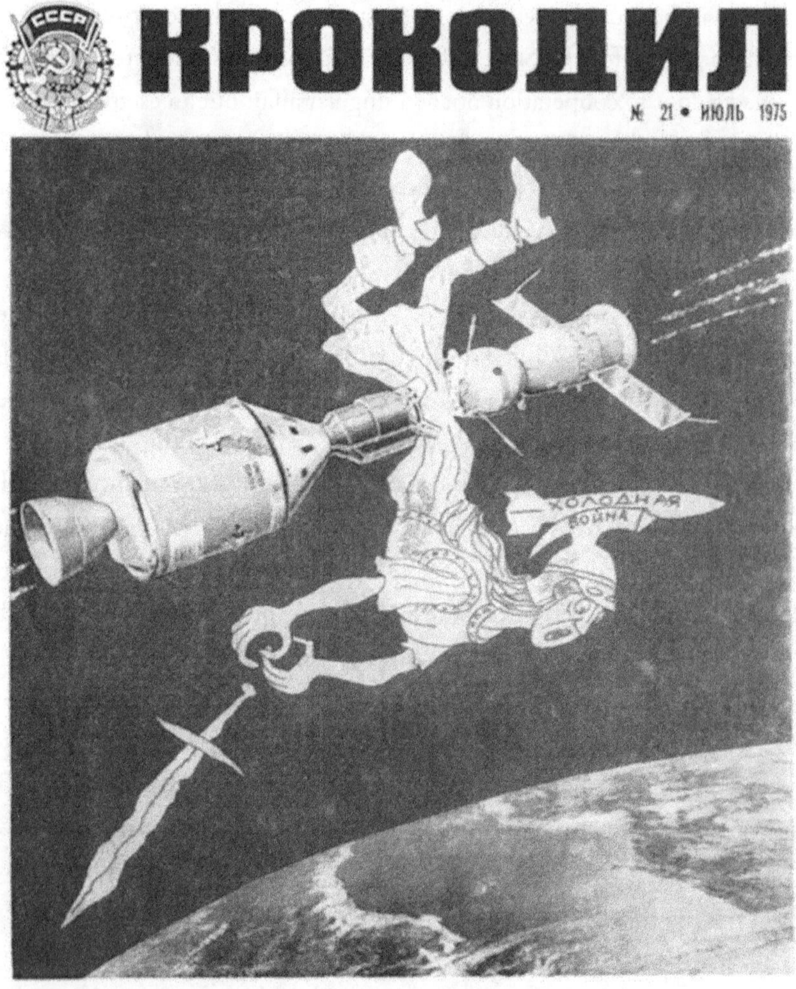

Figure 1: *Krokodil*. Accessed at "СОЮЗ-АПОЛЛОН: Маневры на орбите" [Soyuz-Apollo: On-orbit maneuvers], Live Journal. *Source:* https://1500py470.livejournal.com/136067.html

most likely an armed confrontation, and a Hot War, would ensue. Furthermore, the notion that the actual Cold War would cease cannot be sustained because the Cold War character would be free to roam around again once the two spacecraft discontinue their joint operation.

A similar dichotomy and conflicting interpretation are visible in the American rendering of the "Handshake in Space" mission. As mentioned, the *Time* magazine cover did call the event a "Space Spectacular" (fig. 2). Nevertheless, it stopped short of the fantastic representations of cooperation and mythmaking depicted in the Soviet imagery.

The *Time* cover instead features a handshake where each hand is symbolically painted in ideograms associated with the American and Soviet flags. The handshake itself could be seen as a realistic rendering of the actual event and the televised handshake between the astronauts. But, beyond this realism, the cover itself is also suggestive of the multiple ways in which the joint space mission failed to capture the American imagination and instead became a reason to bury collaborative engagements into the realm of invisibility. Take, for example, the complete erasure of the vast cosmic space and the imaginative sense of exploration beyond the boundaries of the Earth featured on the Soviet image. Similarly, there are no hints of technology, science, or spacecraft—all of which form the very essence of the joint enterprise. Indeed, to a viewer not familiar with the Apollo-Soyuz mission, the image of the handshake has no context apart from the title. The handshake itself could be a representation of any type of interaction between the two opposing sides. The image of the handshake is therefore stripped of the entirety of symbolic associations of space as a universal platform and a common stage where interactions are dedicated to humankind. What is more, instead of an aspirational celebration, the red hand of the Soviet counterpart appears sinister, with a somewhat hidden symbol of the hammer and sickle, only made visible because of the positioning of the hand. Indeed, the Soviet symbol is revealed only because of the act of handshake; otherwise, it would have remained invisible and hidden, as if it belongs to the palm of a spy. And if the impression is that the Soviet counterpart should not be trusted, there is also the grip of the hand, revealing nothing but four ominous nails. The Soviet hand enveloping the American flag is not only ominous and secretive but also claw-like and deformed, with one of the fingers is missing. Although presented as an image of cooperation, the *Time* magazine cover is instead suggestive of the inherent distrust and violence associated with the other side.

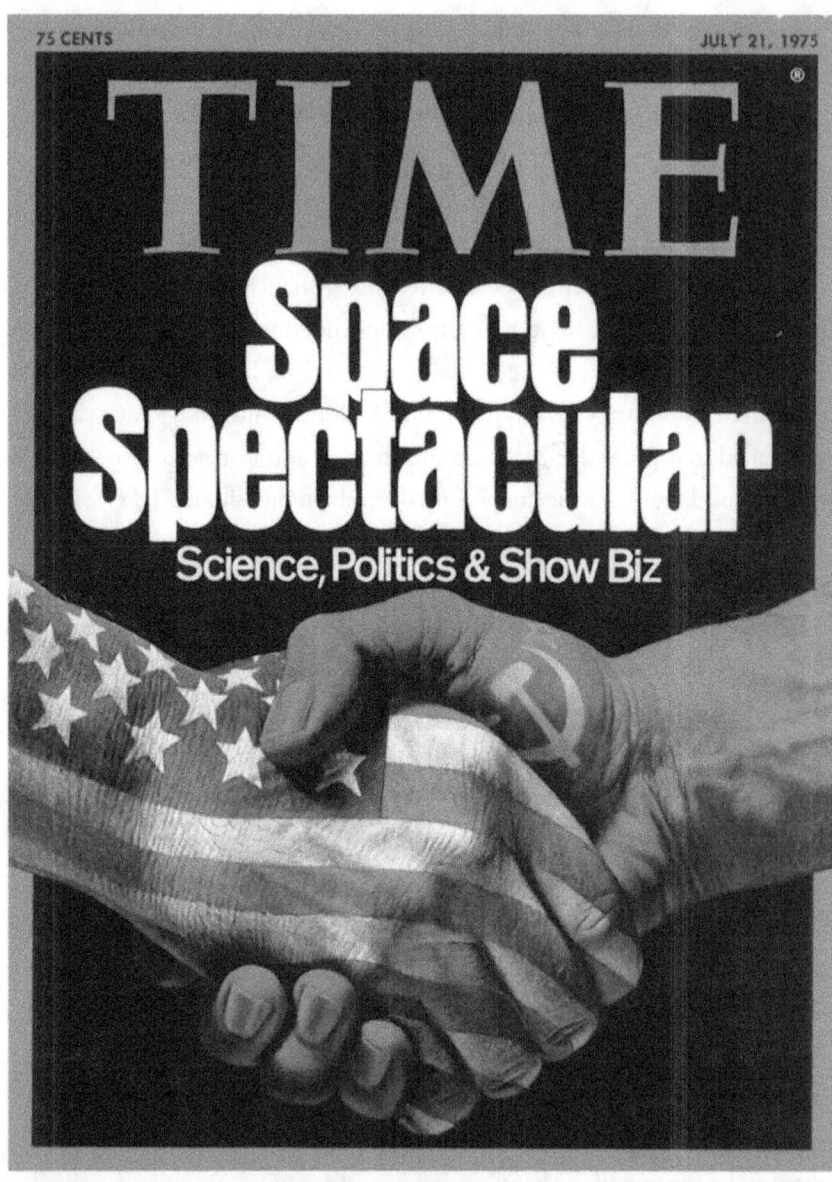

Figure 2: "Space Spectacular," *Time* magazine.
Source: Accessible at https://magazineproject.org/TIMEvault/1975/
1975-07-21/1975-07-21%20page%201.jpg

This interpretation, however extreme, resonated with many of the criticisms of the joint mission, including warnings that the Soviets would use the space cooperation as an opportunity to appropriate superior American technology.[27] The warning that the United States would only stand to lose from such a collaboration was ubiquitous: coming from dissidents, human rights activists, and political refugees. This is also how the very notion of collaborative engagement was stripped of celebratory associations because cooperation itself was seen as dangerous, costly, and ultimately unnecessary.

From the *Time* magazine cover, it appears that the common sense imaginary of collaboration counterintuitively could not exist outside the context of geopolitical tensions, and it was meaningless without such tensions. Even a showbiz-like spectacle was insufficient to bridge the suspicions and mistrust. This was because politically and ideologically the two countries were understood to be so far apart that they could not even muster a common vision for what the cosmic space and the universe beyond planet Earth would look like. Stripped of the evocative imaginary of space, science, and discovery, the cover can be seen as a warning about the challenges of cooperation with the Soviet Union. Ironically bringing visibility to cooperation in the hopes that it would inspire joint action and renewed commitment to finding solutions to challenging global problems seems to have had exactly the opposite effect: in the popular imaginary in the United States, it reaffirmed the suspicions and distrust of the Soviets' intentions.

The Space Race Spectacle

Certainly, it could be argued that the handshake image on the *Time* cover is a reflection of a growing political opposition to space collaboration, especially in light of the exorbitant costs and general distrust of the Soviets. But

---

27. "Space Spectacular," *Time*, July 1975, https://magazineproject.org/TIMEvault/1975/1975-07-21g.

the image also suggests there is something in the imaginary of collaboration itself that is inadequate. In particular, it doesn't appear as if the handshake (as a symbol of collaboration) will lead to anything. Unlike Soviet representations of a comical squeeze of the very cause of confrontation, the handshake image—set against a dark, ominous background—is hardly evocative, impossible to associate with the affective excitement of joint discovery or achievement of improbable goals.

The latter point becomes even more apparent when the handshake cover is compared with another *Time* magazine cover, known as the "Race for the Moon" (fig. 3).

Set in much lighter tones, featuring an unreservedly fantastical rendering of American and Soviet astronauts running toward the Moon, the cover immediately generates a sense of excitement and expectation. There is brightness and energy to the image coming from the two astronauts running toward the moon and the blue background suggestive of the infinity of the universe yet to be discovered. The image also brings excitement because the two figures are so close in their race, like two athletes making a final push toward the finish line. In a final jostle to victory, it is also very much apparent that the American astronaut is about to reach the Moon first.

The contrast between the two *Time* magazine covers, "Space Spectacular" and "Race for the Moon," suggests that the very rendering of space is not informed by the notion of cosmic universe itself but instead exists as part of a larger ideological framework. In the context of the Cold War, this framework positions competition as a meaningful, inspiring, and exciting undertaking. Simultaneously, it reduces cooperation to untold pathways to treachery and deceit. Whereas cooperation appears ominous and leaves the American counterpart vulnerable and unprepared, the image of competition suggests assurance and unlimited ability. According to this image, no one has the capacity to hold and restrict the American running toward the ultimate goal—the conquest of the Moon. This therefore adds to the standard political and economic legitimization of competition as the only way to maintain superiority, independence, freedom, and a clear state of mind. The

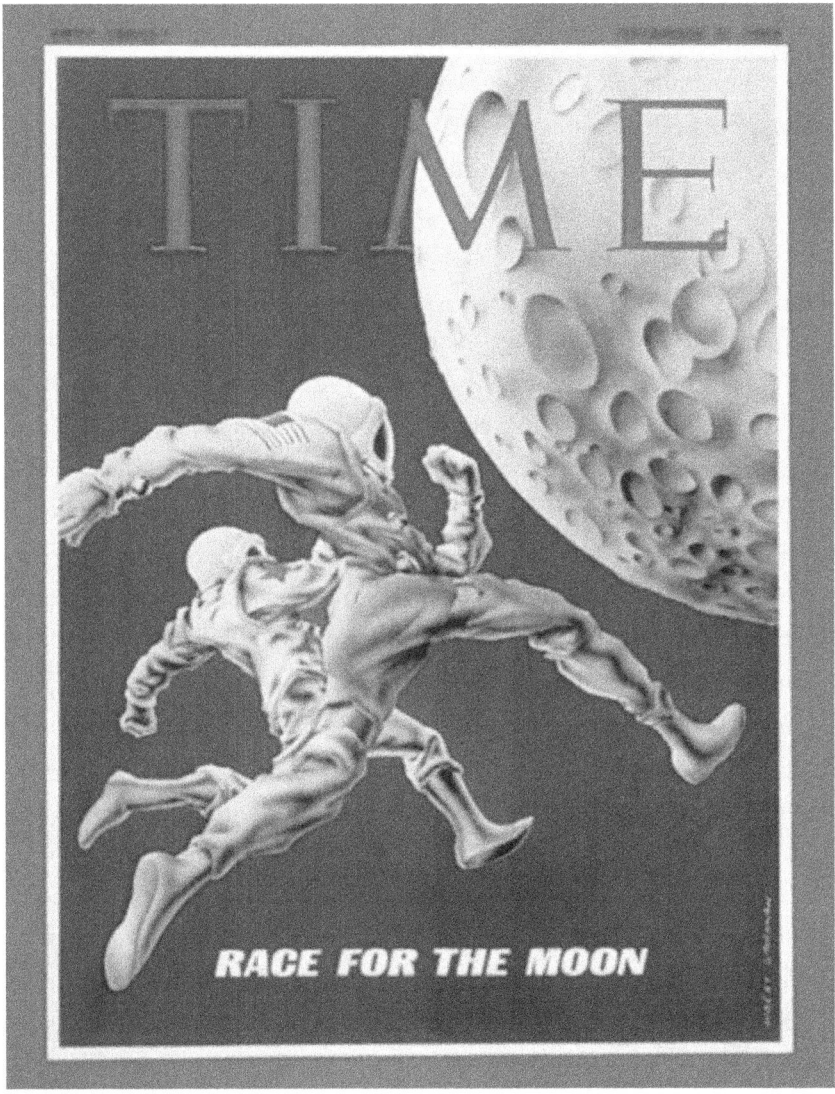

Figure 3: "Race for the Moon," *Time* magazine.
Source: Accessible at http://content.time.com/time/covers/
0,16641,19681206,00.html

image creates a sense that it is only through unhindered competition that new frontiers and the universe could be discovered.

Competition and ideology also feature prominently on the Soviet *Krokodil* covers, albeit with different messaging. Simply stated, because the Soviets were the first to successfully launch into space, the covers reflected the technological and political supremacy of the Soviet Union in space. The message was clear: there is no competition because the USSR had already won the race. This was because there was but one singular presence in the cosmos—the Soviet spaceship—and there could be no competition when the other was simply missing. Consider, for example, the depiction of the smiling sun with a red flaming crown and a Soviet rocket on top of it (fig. 4).

Here, Soviet supremacy is both universal, covering the entire universe, and superhuman because even celestial objects—the sun and the moon—smile approvingly at the arrival of the Soviet rocket. And, whereas the *Time* cover establishes no relation between the humans and celestial objects, in the *Krokodil* imaginary, the Sun itself is turning red (leaning Communist), adorned with a "new diamond" in its crown.

In terms of timing, the two covers—"Race for the Moon" (1968) and "Sun's New Diamond in the Crown" (1959)—both precede and orient common sense perceptions in the decade before the joint Apollo-Soyuz mission and the "Handshake in Space" (1975) took place. As such, they are also suggestive of how cooperation became another cause for distrust and fear instead of a venue for celebration and a hope for a collective decision-making. First, despite the proliferation of institutional declarations and United Nations-led agreements, from the very beginning, space was imagined as an indelible part and natural extension of political, ideological, and technological opposition. Advancement of technology and scientific discovery was not for the purposes of exploration of the universe and benefiting the entirety of humanity but was meant to ensure that celestial objects are "enlisted" according to the two countries' ideological and political preferences.

Second, in framing cosmic space as a new arena for ideological opposition, the common sense terms of engagement were set alongside a

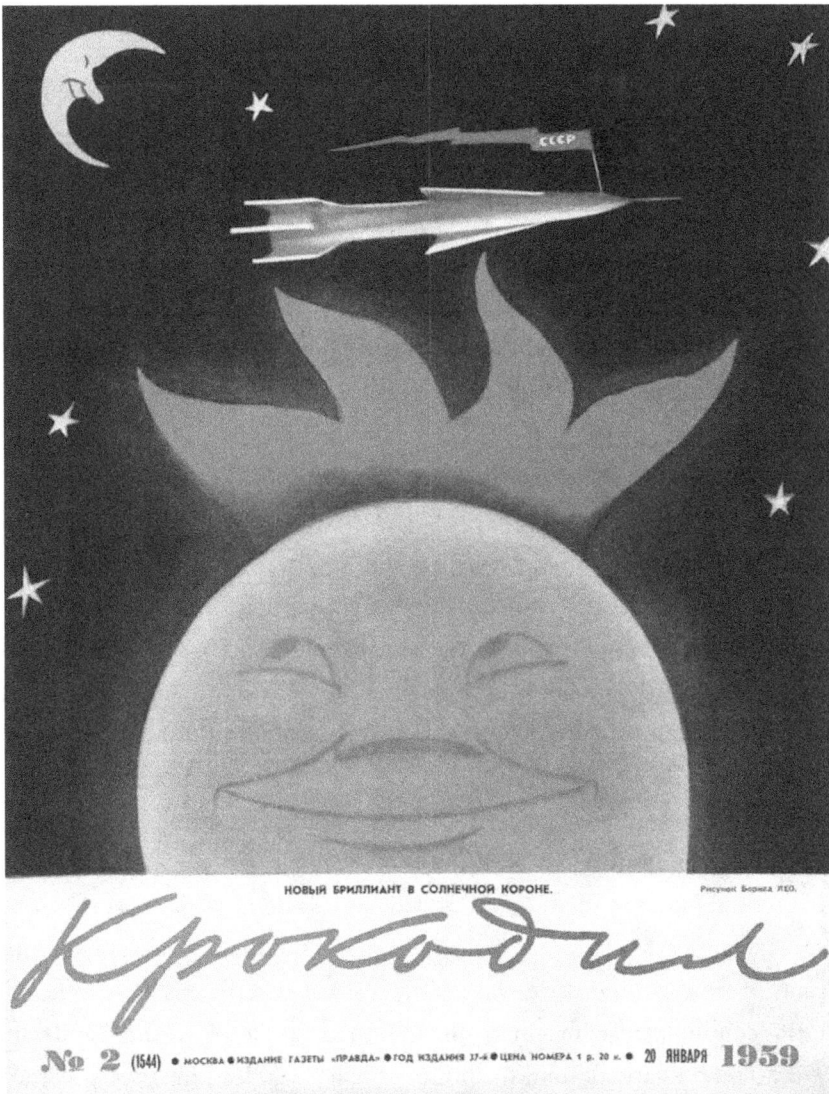

НОВЫЙ БРИЛЛИАНТ В СОЛНЕЧНОЙ КОРОНЕ.

Figure 4: The new diamond in the sun's crown, *Krokodil.*
Source: Accessible at https://coldwar.unc.edu/2018/07/a-new-diamond/

spectacular opposition: the conjuring up of a space race that previously only existed in science fiction. In designating space as an arena for an extraordinary, spectacular competition, however, there was little possibility to maintain such a level of excitement in depicting the tedious, mundane details of cooperation. The image of a handshake (the *Time* cover) could never live up to the imaginary of two people running toward the moon. This, unless the handshake itself could be imbued with a sense of foreboding and treachery. Similarly, the image of two spacecraft docking in space and eliminating a caricatured "Cold War" (the *Krokodil* cover) could hardly compare to an image of Soviet supremacy that extends all the way to the Sun.

Third, unlike the premise of the object-centered theory, whereby cooperation takes place on the basis of a collective, common understanding of an issue of universal concern, there is nothing in the American and Soviet images that would suggest a comparable conception of cosmic space, the goals of venturing into space, or the actual deployment of science and technology for the achievement of such goals. On the *Time* magazine covers, the most prominent visual reference to space comes from a partial image of the moon. Beyond that, space looks like a black abyss or a blue sky. The *Krokodil* covers, although slightly more enhanced, also stop short of presenting a meaningful imaginary of space and include caricatures of the sun and the moon. None of these covers provide any sense of what engagement in space might occasion beyond political references to ideological competition, hints of violence, and the excitement of a space race. To this end, there can be no common sense imaginary on what cooperation would entail and why it is even necessary. Beyond the ominous handshake and the two spacecraft seemingly crashing above the Earth, there appears to be very little meaning to cooperation.

The aesthetics of space cooperation rendered in popular magazines during the Cold War create an impression of cooperation as something invisible or, at best, disorienting. As Ahmed argues, disorientation can occur for a variety of reasons but invariably produces a sense of being lost. To some, such a sense can be exhilarating and trigger reorientation and repositioning of perceptions and understandings. This is also how, on the basis of visual imaginaries, vague, abstract notions of cooperation in space could be discarded for something more "sensible"—such as the excitement of competition. Consequently, unlike the language of the United Nations treaties and political leaders of the time; the imagery associated with space signals how even the vast universe beyond the Earth's orbit might only acquire meaning through the prism of Washington's and Moscow's opposing ideologies and political ambitions.

The Aesthetics of Climate Cooperation

Although dangerously close, the relationship between the United States and China today does not (yet) carry the full ideological weight of the twentieth-century Cold War. However, the aesthetics of climate cooperation already shares resemblances to the Cold War imaginary of space cooperation, therefore also driving a particular type of common sense orientation toward the issue. Just as with the space program, three features of popular images create an impression of an affective incompatibility between the objective for climate cooperation and underlying differences of how the two countries' interactions are seen and thought of. As with the space cooperation described above, these features are the missing universal object, emphasis on competition, and a sense of distrust.

In Chinese periodicals, the notion of climate cooperation rarely appears as part of the popular imaginary. In the few instances where cooperation does feature alongside commentaries on climate change, the overall setting is directly reminiscent of competition. This was the case with a now-discontinued publication in *Duo Wei Online News*, where interactions between the United States and China on climate were strictly presented as an ongoing tussle between the two countries (fig. 5).[28] Indeed, the fact that the *Duo Wei* article and images are now censored and untraceable suggest the very notion of cooperation should not be part of the popular imaginary.

It is nevertheless worth exploring some of the deleted images, as they clearly speak to the dichotomy between cooperation-competition where a common global issue is concerned. One of the images, for example, features

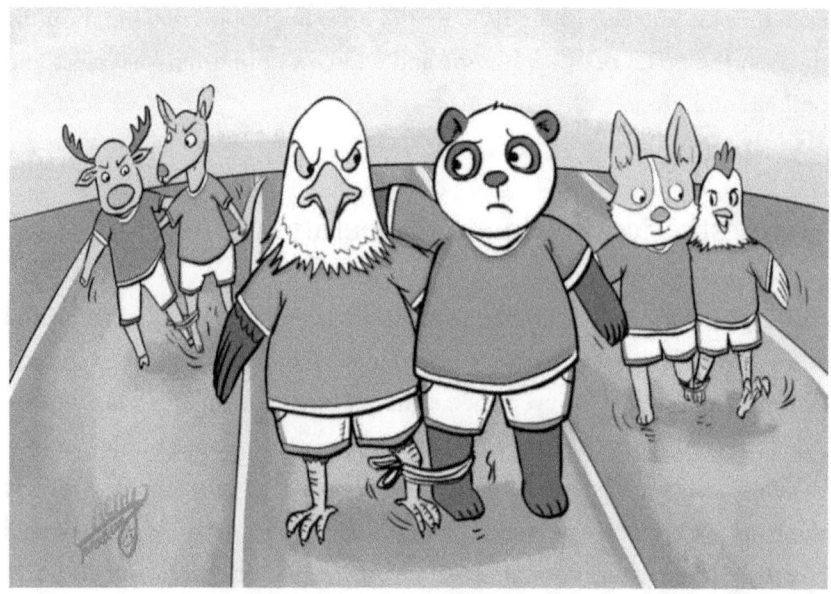

Figure 5: Climate change cooperation, *Duo Wei Online News*. Discontinued. *Source:* Discontinued. Previously accessible at https://www.dwnews. com/%E5%85%A8%E7%90%83/60268398/%E8%AE%AE%E4%B8%96%E5% 8E%85%E4%B8%BA%E4%BB%80%E4%B9%88%E4%B8%8D%E8%96%E5% 8E%85%E4%B8%BA%E4%BB%80%E4%B9%88%E4%B8%8D%E8%83%BD %E7%94%A8%E7%AB%9E%E4%BA%89%E6%9D%A5%E5%AE%9A%E4% B9%89%E4%B8%AD%E7%BE%8E%E5%85%B3%E7%B3%BB

28. C. Kejin, "Why U.S. China Relations Can't Be Defined by Competition," *Duo Wei Online News*, 2021, https://www.dwnews.com/%E5%85%A8%E7%90%83/60268398/ %E8%AE%AE%E4%B8%96%E5%8E%85%E4%B8%BA%E4%BB% 80%E4%B9%88%E4%B8%8D%E8%83%BD%E7%94%A8%E7%AB%9E%E4%B A%89%E6%9D%A5%E5%AE%9A%E4%B9%89%E4%B8%AD%E7%BE%8E%E 5%85%B3%E7%B3%BB..

a depiction of a panda (China) and an eagle (United States) forced to cooperate in order to be part of a competitive event. Judging from the facial expressions and postures, the cooperation arrangement between the two—each participant's leg tied onto the leg of the other—is not going smoothly. Instead of excitement and determination to charge ahead, the two participants appear stuck on how to even perform in unison and share a common space. Each looks suspiciously at the other, considering perhaps how to untie the rope that binds them and run away. Their mutual discomfort is also visible in the awkward hug—the panda is on the verge of twisting its arm while attempting to stay close to the eagle and the eagle is quite uncomfortable with the forced embrace. The image is both a reminder and a stark contrast with the aforementioned "Race for the Moon." In the *Times* depiction, each participant is singularly engaged in the goal of reaching the moon and, as such, determined and focused on the bigger objective at hand. In the *Duo Wei* version, framing cooperation as a necessary part of engagement exposes the limited possibility for achieving anything at all. The two sides are in a state of limbo, stuck between nonconfrontation and noncooperation.[29]

Even the notion of a common global problem that necessitates cooperation would not suffice. In a setting reminiscent of the missing imaginary of universal cosmic space, the panda and eagle image provides no understanding of climate, the universality of the issue, or what exactly cooperation on climate might entail. The conundrum is made even more explicit in China's official news outlet, *Global Times*, where climate cooperation, and not even climate itself, is just another piece on a chess board (fig. 6).[30]

Here, the two countries' have their own "Climate Cooperation" bishop, and the US bishop is already eliminated. The explicit suggestion is that

29. A direct allusion to Anthony Blinken's assertion that the US relationship with China "will be competitive when it should be, collaborative when it can be, and adversarial when it must be." See "A Foreign Policy for the American People," US Department of State, 2021, https://www.state.gov/a-foreign-policy-for-the-american-people/.

30. "China Deserves Praise for Difficult Climate Pledges," *Global Times*, November 2021, https://www.globaltimes.cn/page/202111/1237936.shtml.

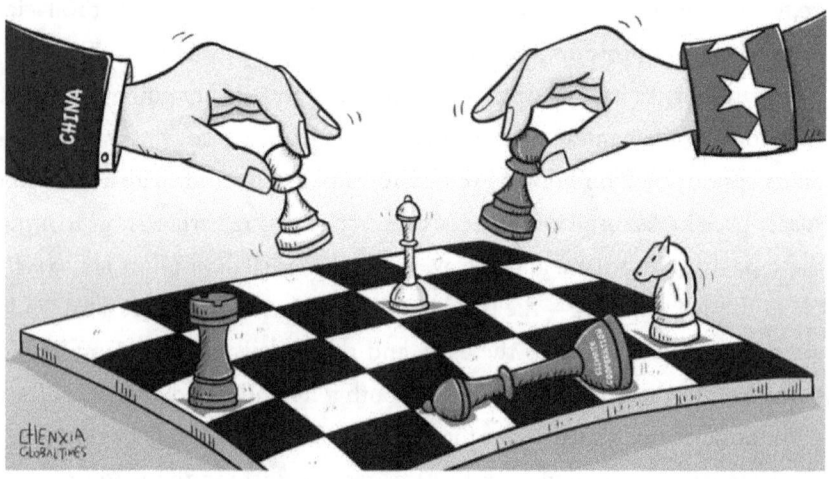

Figure 6: Climate chess, *Global Times*.
Source: Accessible at https://www.globaltimes.cn/page/202111/1237936.shtml

there was never much climate cooperation on Washington's agenda—the bishop has fallen on its own, without any specific move necessary for it to be eliminated. At the same time, the Chinese and American players appear to contend over the fate of the white bishop (i.e., the "Climate Cooperation" piece that Beijing holds). Implicit in the gesture is a sense that the United States is now trying to undermine the climate agenda China might have in place. Even for this reason alone, Washington's agenda on climate is not to be trusted. As was the case with space exploration, a common sense perception of climate cooperation remains fixated on the political opposition and discrepancies between the two countries rather than the actual mechanics of the issue itself. There seems to be a vicious cycle: climate change, and the various interpretations of its meaning, does not exist until the United States and China find a meaningful way to engage collectively. Yet, a collective engagement on something that remains largely undefined is bound to have limited legitimacy or social approval. This impression is further reaffirmed by the chess board setup: with both kings missing, the entire undertaking appears to be a sham. Aligned with the Chinese political leaders' continued insistence that climate cooperation cannot be viewed separately from the overall relationship between the two countries,[31] the image reasserts a common sense impression that the game of cooperation cannot be played with only some pieces, pretending that others are somehow irrelevant to the setup.

The theme of visibility-invisibility, presences-absences, is also central to the *Time* magazine "Last Call" rendering of climate change (fig. 7).

The cover reflects the bizarre configuration during the COP26 climate meeting in Glasgow where many world leaders and delegates were unable to attend in person due to the ongoing COVID-19 pandemic. The "Last Call" reference itself could be interpreted in multiple ways. For example, in a since-deleted "Last Call for Climate" tweet, *Time* magazine staffers suggested the Glasgow event was the last opportunity to commit to meaningful joint action to prevent climate change. However, "last call" could also be understood as an attempt to summon those who are yet to come to the hallways of negotiation and a prompt that they reaffirm their commitment to united action. Such a call is especially directed toward the missing Chinese leader, Xi Jinping. His name placard is placed in between the respective leaders of the United States and the European Union, Joe Biden and Ursula von der Leyen, who are just seen waiting. Featured at the very front, the latter seem to be in charge of the "last call." They are the leading proponents (along with activist Greta Thunberg, the president of Nigeria, and the prime minister of India) of the need for collective action. Seemingly, those who are missing, who fail to join in global climate cooperation efforts, are the ones who are still bent on competition while the rest of the world is on the verge of burning, drowning, and freezing—all at the same time. The image therefore is suggestive of the failure of the Chinese side to show up at the most

---

31. J. Shi, "Climate Crisis: China's All-or-Nothing Stand on Talks Leaves John Kerry Cornered," *South China Morning Post*, September 7, 2021, https://www.scmp.com/news/china/diplomacy/article/3147854/climate-crisis-chinas-all-or-nothing-stand-talks-leaves-john.

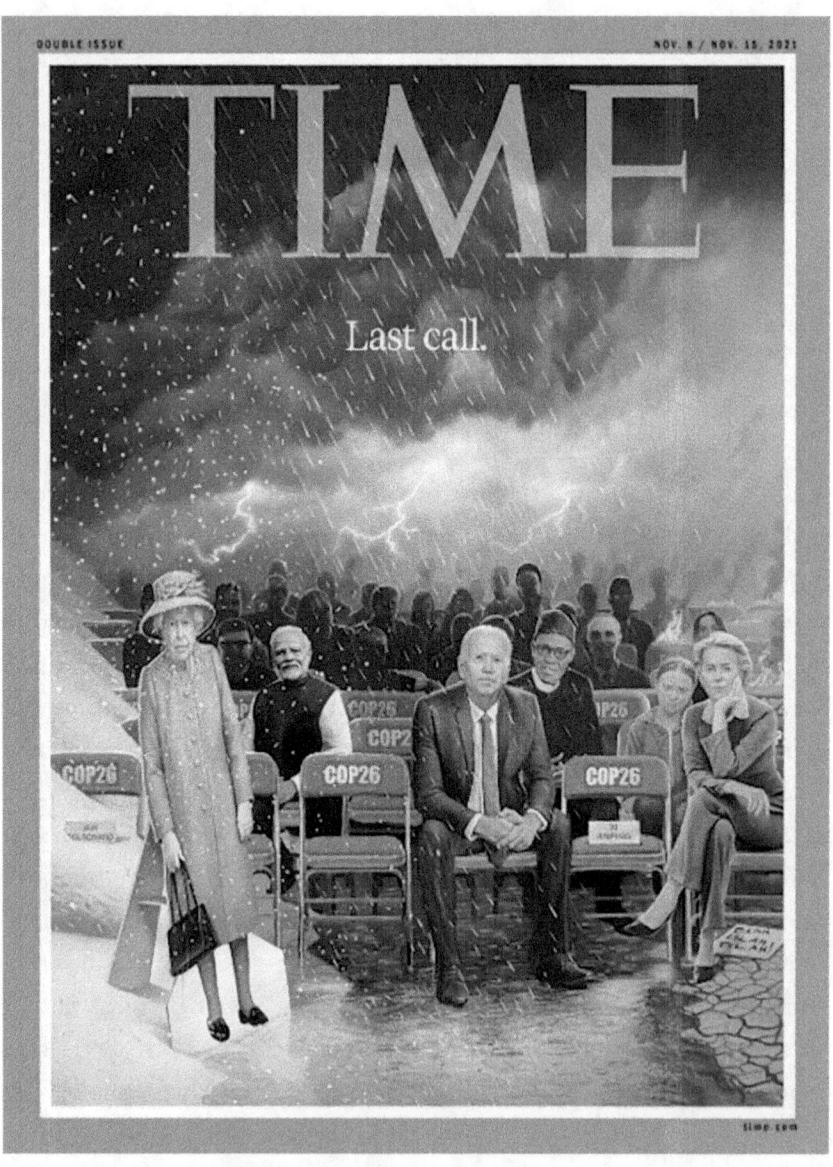

Figure 7: "Last call," *Time* magazine.
Source: Accessible at https://time.com/6109403/cop26-summit-agenda

critical juncture of time and place. The "last call" is the last opportunity for cooperation, where the entire world is waiting for China to deliver on its alleged commitment. The treachery and unmet expectations are alluded to, both visually (the missing presence) and literally (last call—last supper).

It is worth noting, however, that unlike other depictions, the "Last Call" image presents a unique, complex imaginary of climate change. In a rare depiction of its complexity, climate change is rendered not as one singular phenomenon, or an image of a green Earth, but as a combination of cataclysmic events happening all at once. On the cover, some of the assembly chairs are on fire, referencing the devastating fires across many parts of the world. At the same time, at the other end of the assembly hall, the chairs are covered in deep snow. Even Biden and von der Leyen are not spared: they appear to be standing in a space that is simultaneously in danger of flooding and desertification.

Certainly, extreme weather events are considered to be one of the main effects of climate change, yet this cataclysmic background can be also deceiving. As many climate deniers might argue, fires and draughts are just normal weather fluctuations, nothing extraordinary in the overall pattern of a constantly changing global environment. As a typical saying in climate change denial goes, "the climate always changes."[32] Moreover, Biden and von der Leyen themselves appear oblivious to the cataclysmic environment surrounding them: either because they are used to it or because they do not want to acknowledge the severity and catastrophic impact of weather fluctuations. Indeed, a more cynical interpretation of the image would be that world leaders, irrespective of whether they are present or absent from an event, remain impervious and unfazed by the apocalyptic conditions unfolding right in front of them. Their serene faces suggest a perfect disconnect from the cataclysmic combination of snow, rain, fire, and wind surrounding them. Instead, they sit calmly, hands or legs crossed, staring at a void ahead, expectantly waiting.

---

32. See, for example, Tucker Carlson's climate debates, such as "Tucker vs. Bill Nye the Science Guy," Fox News, YouTube, https://www.youtube.com/watch?v=qN5L2q6hfWo.

Such interpretation leaves an impression that whereas the world is on the verge of falling apart, major political leaders can only sit and wait in oblivion. As such, the political message concerning the missing Chinese leader could also be lost in the havoc of the cataclysmic surroundings and general inaction of those present. Getting one more person to sit on a chair alongside Biden and von der Leyen is hardly *the* solution to the apocalypse enveloping the image. Ultimately, it remains unclear what the world leaders are meant to do with respect to the dramatic weather conditions and, in particular, how cooperation is linked to these extreme events. Similar to the issue of space exploration, there is nothing in the imagery on climate change that points to the value of cooperation.

Conclusion

What does cooperation between rival superpowers look like? Do global issues have the capacity to rise above the geopolitics of the day and trigger alignment between rival powers? This article argued that the Cold War joint space exploration program between the United States and the USSR provides a lesson on the limits of cooperation. These limits, I posited, are not only a matter of power preferences, institutional differences, material disincentives, or even a consequence of a tendency for free-riding. Rather, they are also the result of incompatible common sense perceptions. Such perceptions are formulated and driven by overarching mistrust and fear of the strategic motivations and hidden purposes of the other side. Cooperation, even if institutionally viable, is constrained due to a lack of popular endorsement and legitimacy.

To explore how common sense perceptions structure a sense of legitimacy, the article turned to the politics of aesthetics and interrogated the visual renderings of space interactions in the United States and the USSR. Using a phenomenological approach and drawing on images widely circulated in popular magazines, I showed how the imaginary of cooperation itself can become the source of fear and distrust. Images of a handshake or an arm across the shoulder do not necessarily signal agreement and trust; instead, they can become the source of disorientation, further contributing to inherent predisposition to shun cooperation. This is also how a viable space collaboration remained out of public sight: this even though it continued for many decades, and well beyond the collapse of the USSR.

As political differences and sense of ideological incompatibility envelop global capitals today, the Cold War imaginary of the space program can serve as both critique and inspiration for present day attempts at cooperation on global issues such as climate change. Four insights from the aesthetics of cooperation suggest a good starting point in heeding the lessons of the past.

First, despite the existence of a global framework and considerable political support at the highest level of government, the Cold War imaginary suggests there is no common understanding and therefore no legitimate grounds for cooperation on a universal issue. Relatedly, the notion that a universal issue would inspire a common approach "beyond politics" does not hold. The problem is not simply in the mechanics of finding a common approach. It is also in agreeing on the nature of universals and their validity.[33] Turning to the politics of aesthetics is one way of showing that there is no common sense understanding of universals such as "space" or "climate change." Visual renderings of space across different popular magazines revealed divergent conceptualizations of universality and what an engagement with a "universal" issue might entail. It is notable, for example, that images of cooperation in space did not reference joint research or collaborative development of technology (as was actually the case). Instead, the popular imaginary of cooperation in space was limited to either a vision about the end of the Cold War or a handshake. At the same time, the entire process of technological cooperation and attempts to make the two programs interoperable. Today, climate change is similarly visualized differently

---

33. See also A. L. Tsing, *Friction: An Ethnography of Global Connection* (Princeton, NJ: Princeton University Press, 2004).

by different sides, with representations ranging from a chessboard juggle to an apocalyptic world with multiple weather calamities happening all at once.

Second, and related, in a world of extreme geopolitical tensions and bipolarity, it appears unrealistic to expect cooperation between hostile superpowers to resolve a widespread sense of animosity and distrust. At the very height of space cooperation, at the moment of a spectacular joint mission, the overarching common sense understanding was that cooperation was a costly and unnecessary enterprise that would only benefit and bring prestige to the enemy. There was nothing inherent in the nature of space itself that made it more conducive to collaborative action. To the contrary, the common sense understanding of how humans relate to space remained vastly different and collaboration could not be legitimized because of entirely different aspirations associated with space.

Third, while competition is not seen as conducive to a peaceful relationship, the space programs and the joint attempts at expanding technical capabilities and reaching the moon point to a phenomenon of "competitive cooperation." While visibly, in the public space, the two sides were seen as competing against each other, the competition itself pushed both countries to dedicate the resources, advance training opportunities, and supply the necessary conditions for technological innovation. All of this was possible because there was another side to partake in the competition and keep the race going. At the same time, the competition was justified and legitimated as an assertion of power and technological superiority. One lesson for climate change engagement could therefore be that it is competition, rather than cooperation, that might inspire the necessary level of innovation and technological breakthroughs needed to prevent a climate catastrophe.

A final lesson of Cold War cooperation relates to the overarching power of cognition and orientation of common sense understandings. At times of extreme ideological competition, there is an expectation to see deception and duplicity irrespective of government agendas, institutional arrangements, and celebratory media portrayals. Furthermore, in a setting of extreme politicization and ideological competition, the public thirst for spectacular excitement comes not from a dubious handshake symbolizing cooperation but from the imagery of fantastic, superhuman competition. As such, suggesting a cooperative engagement around something nebulous and under-defined, such as space exploration, only extends the sense of affective dissonance and distrust. In a world edging closely to a new bipolar hostility, the Cold War lesson on cooperation is a warning on the power of affective disorientation that has the ability to distort objectives, undermine legitimacy, erode trust, and erase the very imaginary of collective action. In a manner similar to the erasure of the collective memory of space cooperation during the Cold War, today there appears to be little ground left on which an imaginary of climate cooperation might prevail over the overwhelming thrust toward competition.

Acknowledgments

I would like to thank Kenneth Paul Tan for the invitation to contribute and his constructive feedback throughout the entire process. I am greatly indebted to Iliyana Nalbantova and Gao Peng for their meticulous search of Soviet and Chinese images and comments on earlier drafts.

Book Reviews

Through Space and Time

Review of *The Odyssey of Communism: Visual Narratives, Memory and Culture* edited by Michaela Praisler and Oana-Celia Gheorghiu, Cambridge Scholars Publishing, 2021

ISABEL GALWEY

Keywords: communism, cinema, socialist cinema, Soviet cinema, Chinese cinema, Eastern Bloc, Yugoslavian cinema, Romanian cinema, Polish cinema

The Odyssey of Communism: Visual Narratives, Memory and Culture traces the intellectual history of Communist visual culture from Ceaușescu's Romania to Mao's China, as well as other countries in the Eastern Bloc and the non-Communist world. Its editors are Michaela Praisler and Oana-Celia Gheorghiu of the University of Galati, Romania.

Based on the proceedings of the 2019 Romanian conference "Thirty Years since the Fall of Communism: Visual Narratives, Memory and Culture," the volume features a polyphonic range of works by authors based in Turkey, Poland, Ukraine, Serbia, Romania, Hong Kong, the Netherlands, and the United States. *The Odyssey of Communism* revisits the recent history of Communism through the lens of cinematic culture. The book aims to present a multidisciplinary perspective on Communist film, within a broad definition that includes both film texts produced in (former or presently) Communist countries and in the West. This book forms a valuable

contribution to the study of Communist cinema in English-language academia, drawing together several different areas of scholarship—including
film theory, history, and cultural studies—in one engaging volume.

Taking the form of an episodic voyage of discovery, *The Odyssey of Communism* is divided into three parts, plus a coda. The first, entitled "Hades:
The Red Turns to Black," deals with films that directly or indirectly depict
the darkest aspects of the "Communist inferno," from the subversively critical Yugoslavian "Black Wave" film movement to the carceral spaces explored
through Tarkovsky's *Stalker* (1979) and the 1993 film adaptation of Marin
Preda's novel *Most Beloved of Earthlings* (1980).

Evoking Dante, the authors of this first and longest section use their film
texts to reexamine some of the most difficult and painful episodes in Communist history, through close readings that at times employ philosophical or
psychoanalytical approaches. For example, chapter 5 ("Abortion in Ceauşescu's Era: From Personal Drama to Social Problem") deals with the psychological implications and fraught gender politics of Romania's 1960s abortion
ban, "one of the most abusive and repressive measures for population growth
in history."[1] This chapter stands out particularly for its thoughtful comparison between two films, which depict Romanian women during this period,
exploring the ways in which different characters respond psychologically to
the traumatic political events that intrude into their lives and bodies. The
films—*Postcards of Wild Flowers* (1975) and *4 Months, 3 Weeks and 2 Days*
(2007)—provide complementary perspectives, running the gauntlet from
tragedy to absurdity. In chapter 1 ("The Repercussions of Finding a Voice:
Silent Wedding"), Alexandru and Michaela Praisler also explore the interplay
of farce and tragedy in *Silent Wedding* (2008); their work contributes to the
rapidly growing intersection of cinema and sound studies. Due to a ban on
all celebrations following the death of Stalin, a couple in a small Romanian

---

1. Felicia Cordoneanu, "Abortion in Ceauşescu's Era: From Personal Drama to Social Problem," in *The Odyssey of Communism, Visual Narratives, Memory and Culture* (Newcastle
   Upon Tyne: Cambridge Scholars Publishing, 2021), 56.

village hold their wedding celebration at night, in complete silence: "Staged and silent, farcical and pathetic, [it] is the epitome of debasement, allowing the comic to dissolve into the tragic."[2] The authors analyze the creative use of silence in the film: at times, the lack of dialogue alludes to cinema's silent origins to intensify the farcical physical comedy. However, the film's central themes of silence as a form of subversion or resistance also emphasize the complex relationship between speech, silence, and censorship.

The second part of the book, "Lotus Eaters. Propaganda, Intoxication and Complacency," focuses on the implications of Communist cinema as propaganda. Chapter 6 ("Six Decades of Spring: Refashioning the Soviet Industrial Myth in *Spring on Zarechnaya Street* (1956)" by Daria Moskvitina and Bohdan Korneliuk) traces the popular and critical receptions of the Soviet Ukranian classic *Spring on Zarechnaya Street* (1956). This chapter stands out for its exceptionally clear and engaging analysis of the enduring cultural legacy of *Spring on Zarechnaya Street* in the Ukranian city of Zapor-izhzhia. The film is significant both as a relatable urban romance and as an exploration of the tensions between representations of intellectuals and industrial workers in Soviet popular culture. By contrast, in the following chapter ("Anecdotal Takes on Social History: Tales from the Golden Age Told and Re-Told" by Oana-Celia Gheorghiu), the author focuses on satire rather than nostalgia. She compares a 1975 Romanian propaganda film, *The Freshmen's Autumn*, with a satirical omnibus film series produced in the early 2000s, *Tales from a Golden Age*, which brings to the fore the absurdist elements that were buried in the former work.

The final two chapters of this section ("Something Is Rotten in Film Propaganda: Ideological Games and Self/Other Representations in *Comrade Detective* (2017)" by Gabriela Iuliana Colipcă-Ciobanu and "From Holly-wood to the Soviet Model: Building a Socialist Chinese Cinema" by Ying Zhu) considerably widen the volume's scope. Western representations of

---

2. Alexandru and Michaela Praisler, "The Repercussions of Finding a Voice: *Silent Wedding*," in *The Odyssey of Communism*, 12

Communism come to the fore in a discussion of the show-within-a-show *Comrade Detective*, which was created by a Western production team for Amazon Prime to resemble a piece of 1980s Romanian propaganda. This nested structure of pastiche brings to mind Kristen Daly's assertion that contemporary cinema, or "Cinema 3.0," takes on deliberately game-like characteristics, "based on interactivity, play, searching, and nonobvious relationships."[3] While, like *Tales from a Golden Age*, there are satirical elements to *Comrade Detective*, the authors argue that ultimately it is "a deeply dialogic text . . . whose point may not be so much to produce the caricature of an antagonistic way of thinking but to question the validity of any dominant ideology."[4] The final chapter shifts eastwards, as Ying Zhu uses a more historical perspective to insightfully chart the fortunes of Hollywood and Soviet cinema in China directly pre- and post-1949.

The third part of the book deals with the nostalgic recollections of the Communist era through recent reflective and retrospective films produced by a younger generation of filmmakers. Kaixuan Yao's chapter, "Schizo-Historicising with the Body: Representations of Dance in Recent Cold War Nostalgia Films," is particularly notable for her insightful discussion of the role played by dance in Cold War "nostalgia films," in which she analyzes the aesthetic and intellectual appeal of the Cold War from a contemporary perspective: "The beautifications or, at least, elevations of the Cold War memories in comparison with the present point to the use of the past as an alleviating, though self-deceiving discourse."[5] For example, in her discussion of *White Crow* (2018), which tells the story of Soviet dancer Rudolf Nureyev's defection in Paris, she emphasizes the importance of Soviet material

3. Kristen Daly, "Cinema 3.0: The Interactive-Image," *Cinema Journal* 50, no. 1 (2010): 81–98, 83
4. Gabriela Iuliana Colipcă-Ciobanu, "Something Is Rotten in Film Propaganda: Ideological Games and Self/Other Representations in Comrade Detective (2017)," in *The Odyssey of Communism*, 136
5. Kaixuan Yao, "Schizo-Historicising with the Body: Representations of Dance in Recent Cold War Nostalgia Films," in *The Odyssey of Communism*, 165

culture and a sense of place for creating a mood of nostalgia in the film, a sense of home that makes Nureyev's eventual decision to leave for the West more meaningful. She also draws a pertinent link between films like *White Crow* and *The Shape of Water*, which feature dance explicitly on-screen, and balletic, elaborately choreographed action films such as *Atomic Blonde*. Eleni Varmazi's "Revisiting Germany's Communist Past after the Fall of the Berlin Wall: *The Lives of Others* (2006)" examines an attempt at post-Communist reconciliation in the 2006 drama *The Lives of Others*, in which a (fictional) Stasi officer defies his superiors to protect an idealistic intellectual upon whom he has been requested to spy. Varmazi argues that although the film received criticism for its perceived inauthenticity, painting perpetrators of surveillance in a sympathetic light, it is redeemed by its sensitive portrayal of individual struggles of the sort seldom captured by the broader strokes of history.

Chapter 12 ("To Make Ends Meet: Downplayed Struggle During the 1989 Polish Transition" by Olga Szmidt) presents another challenge to conventional Western historiographies of the Cold War era and the monolithic Eastern bloc. Szmidt explores the diversity of political, social, and economic realities in Poland during the *transformacja* (Polish for "transition" or "transformation") period before and after 1989. Finally, chapter 13 ("The Image of the Mayor in Communist and Postcommunist Romanian Filmography" by Monica Manolachi) leads the reader back to Romanian cinema, where the volume began. Chapter 13 focuses on the ambivalent figure of the mayor—whether they are a hero, rogue, or villain—in Communist and post-Communist cinema. Particularly interesting is the discussion of *A Bird's-eye View upon the City* (1975). Its narrative, centered around a capable female mayor, combines "melancholic and poetic" storytelling with "subtle satire," offering more nuance than the socialist heroes who populated screens of earlier decades.[6] Through the evolution of these fictional mayors, Monica Manolachi charts the trajectory of Romanian cinema across the decades.

---

6. Monica Manolachi, "The Image of the Mayor in Communist and Postcommunist Romanian Filmography," in *The Odyssey of Communism*, 219

The final part of the book, "Coda: Cyclops," consists of a single chapter ("'This Is (Not) A Fairy Tale': Documenting the Orsinian Revolution" by Gabriela Debita). It reopens the cycle of the *Odyssey*, probing the boundaries between cinema and other art forms through its analysis of Ursula Le Guin's short story "Unlocking the Air," which is set in the imagined socialist state of "Orsinia" during its fictional collapse. Debita highlights the story's fragmentary structure: she argues that Le Guin's style mimics both the montage of cinematic editing and the fast-moving kaleidoscope of late twentieth-century mass media. The story's cinematic qualities, Debita suggests, evoke the zeitgeist of the end of the Cold War in Central and Eastern Europe and the self-conscious making of history through camera lenses. Using "Unlocking the Air" as a starting point, Debita interrogates the dynamic and sometimes problematic relations between Communism as it is remembered and reimagined in the West and how it was experienced by those who lived through it. She also points out that some of the heady promises of the era directly following the end of the Cold War have not come to pass, complicating narratives of this period still further. Although Le Guin reminds her readers that "this is not a fairy tale," Debita suggests that Orsinia still offers readers a moment of uncertain, almost magical optimism, boldly declaring the start of a new journey. Her favorable reading of Le Guin's short story, like Eleni Varmazi's interpretation of *The Lives of Others*, focuses on visual narratives as a space for reconciliation and imagination, and for making sense of the recent past.

The Odyssey of Communism's subject matter is highly diverse. In their foreword, Praisler and Gheorghiu emphasize the lightness of their editorial touch, particularly regarding the ideological implications of any chapter in the volume. *The Odyssey of Communism* covers an ambitious amount of ground, with subject matter stretching across time and space, covering more than half a century's worth of history in a variety of geographical regions. While this diversity is admirable, it does at times stretch the limits of the book's four-part structure, which can seem arbitrary in its divisions. (For example, the contemporary satire *Tales from a Golden Age* is discussed in chapters 7 and

13, falling in the second and third parts, respectively.) Conversely, this can leave the connections between chapters somewhat disjointed.

The large scope of the volume also means that the content at times can tend toward the broad-brush or descriptive, as in the case of chapter 9, on the topic of cinema in the early days of the PRC, which covers ground touched upon in other volumes such as Zhang Yingjin's *Chinese National Cinema*.[7] Chapter 12, on the role of the mayor in Romanian cinema, reads at times like a filmography and might have benefited from a few closer analyses. The inclusion of films made in Western countries, turning their lens (or pen) upon Communism from an outsider's perspective—for example, chapters 8, 10, and the coda, which analyze depictions of Communism in Anglophone film, television, and literature—stretches the scope of the volume still further.

However, this epic scale and polyphonic content is also what makes *The Odyssey of Communism* an engaging and compelling book. The very diversity of its subject matter highlights the ideological and historical complexity of Communism, which runs through the book like a thread on Penelope's loom. Its boldly interdisciplinary approach would make it valuable to scholars of film history, film studies, comparative literature, cultural studies, and political history, among others. *The Odyssey of Communism* emphasizes the complex subjectivity of Communist cultural workers and their negotiations with politics, ideology, and identity. In reopening a period of recent history that is fast rigidifying into myth or new forms of propaganda, *The Odyssey of Communism* highlights the myriad ways in which the stories and images of the Cold War period continue to impact global narratives today.

Acknowledgments

Thank you to Dr. Kenny K. K. Ng for the valuable guidance and thoughtful feedback on an earlier version of this review.

---

7. Yingjin Zhang, *Chinese National Cinema*, 1st ed. (London: Routledge, 2004), ch. 6.

Review of *Hollywood in China: Behind the Scenes of the World's Largest Movie Market* by Ying Zhu, New Press, 2022

YONGLI LI

Ying Zhu's *Hollywood in China: Behind the Scenes of the World's Largest Movie Market* is one of the newest additions to the Sino-Hollywood film industry corpus and their entanglements, conflicts, cooperation, and controversies over the past century. Zhu examines the vicissitudes of the two largest film industries in the world by chronicling their histories against prominent global political and cultural changes, from early Republican-era China to the (post-)COVID-19 era. Through archival research, case studies of popular films and film companies, and interviews with media practitioners, Zhu unveils the dynamic relationship between China and US film industries and its ramifications to the global political and cultural hierarchy.

In 2012, many trade journals predicted that China would surpass the United States and become the world's largest film market by 2020. These predictions came to pass, but only amid the pandemic in 2020, when most of the world's cinemas were under serious restrictions. Since the early 2000s, increasing numbers of scholars[1] have shed light on the marketization and

---

1. Michael Curtin, *Playing to the World's Biggest Audience: The Globalization of Chinese Film and TV*, 1st ed. (Berkeley: University of California Press, 2007); Darrel William Davis, "Market and Marketization in the China Film Business," *Cinema Journal* 49, no. 3 (2010): 121–25; Darrel William Davis, "Marketization, Hollywood, Global China," *Modern Chinese Literature and Culture* 26, no. 1 (2014): 191–241; Emilie Yueh-yu Yeh and Darrell William Davis, "Re-Nationalizing China's Film Industry: Case Study on the China Film Group and Film Marketization," *Journal of Chinese Cinemas* 2, no. 1 (January 1, 2008): 37–51;

globalization of contemporary Chinese film and media industries and analyzed the cultural and political ramifications of these changes. As the Chinese market made itself very attractive to foreign investors beginning in the early 2000s, Hollywood explored more ways to access it and explored various approaches to collaboration and coproductions. However, gradually increasing political tensions between China and the United States since 2017 reversed the "honeymoon" period between China and the Hollywood film industry. In the past two decades, scholars[2] have assessed and questioned China's "soft power" and "going-out" strategy and their complications and consequences by focusing on the contemporary China and Hollywood relationship.

Zhu's framework for analyzing the shifting dynamic between China and Hollywood centers on the impacts of local and global political interventions and how these interventions further impacted global political and cultural discourses. Through nine chapters, Zhu moves through historical roots back to the semicolonial period of cosmopolitan Shanghai and documents the most recent moment of global expansions of China and Hollywood amid the COVID-19 pandemic. Written in chronological order, the book can be generally divided into four sections. Chapters 1 and 2 conjure up a path of early Republican-era China to the 1940s postwar Shanghai film industry; chapters 3 and 4 sketch out the socialist period to early Reform-era China; and chapters 5 and 6 detail

---

Ying Zhu, *Chinese Cinema during the Era of Reform: The Ingenuity of the System* (Westport, CT: Praeger, 2003); Ying Zhu and Chris Berry, *TV China* (Bloomington: Combined Academic Publishing, 2009).

2. Aynne Kokas, *Hollywood Made in China* (Berkeley: University of California Press, 2017); Stanley Rosen, "Obstacles to Using Chinese Film to Promote China's Soft Power: Some Evidence from the North American Market," *Journal of Chinese Film Studies* 1, no. 1 (May 1, 2021): 205–21; Wendy Su, *China's Encounter with Global Hollywood: Cultural Policy and the Film Industry, 1994–2013* (Lexington: University Press of Kentucky, 2016); Yiman Wang, *Remaking Chinese Cinema: Through the Prism of Shanghai, Hong Kong, and Hollywood* (Honolulu: University of Hawai'i Press, 2013); Michael Berry, "Chinese Cinema with Hollywood Characteristics, or How *The Karate Kid* Became a Chinese Film," in *The Oxford Handbook of Chinese Cinemas*, ed. Carlos Rojas and Eileen Chow (Oxford and New York: Oxford University Press, 2013); Davis, "Marketization, Hollywood, Global China."

the marketization of the Chinese film industry and its cooperation with Hollywood during the 1990s through 2010. Finally, the last three chapters examine the popular Chinese films at home and "going out," as well as the coupling and decoupling of China and the Hollywood industry from 2010 to 2021.

In the introduction, Zhu revisits the legendary Paris Theater and its romanticized portrayal in the new sensationalist writer Shi Zhecun's short story. Located in the former French Concession of Shanghai, the well-known foreign-invested theater has experienced a series of name changes. From Palais Oriental Theater (1926) to Peacock Oriental Theater (1927), to Paris Theater (1930) to the Huaihai Cinema (1951) in the socialist period and Huaihai (Times) Cinema (1993) in the heyday of real estate speculation in the 1990s, all of which reflected the changing economic, political, and social discourses of the time. The theater stands in for the vicissitudes of Shanghai under various regimes, and its history condenses the cultural and commercial exchange between the Chinese and the world. From the colonial era representing China's modern cosmopolitanism to the intensification of post–Cold War globalization to the decoupling amid deepening political polarizations in the COVID-19 era, Zhu argues that "Hollywood has become . . . more a clearing house for global financial and creative forces gathered under the corporate rubric of the 'Hollywood Way,'" and China has "redeployed the Way with Chinese characteristics."[3]

In the first two chapters, Zhu argues that Shanghai, where Hollywood's global expansion met the local buildup and protection, occupies a central position in the Chinese film industry in the Republican era. Drawing upon archival research, Zhu conveys the prevalence of Hollywood films and their popularity in Shanghai and notes the early reliance of Chinese film productions on foreign (mainly Western) investment, technology, and

---

3. Ying Zhu, *Hollywood in China: Behind the Scenes of the World's Largest Movie Market* (New York: New Press, 2022), 8.

popular genre conventions. From the first Chinese feature film, *The Difficult Couple* 难夫难妻 (dir. Zhang Shichuan 张石川 and Zheng Zhengqiu 郑正秋 1913) to social-realist melodramas and costume dramas in the 1920s, Zhu argues that it was waves of "westernization" under the influence of the May Fourth Movement and sanitization that formed the initial enthusiasms of cinema audiences in China.[4] Zhu also analyzes the political interventions of the governing parties in the Chinese film industry and argues that the film industry's "entanglements with political parties" during the 1920s and 1930s was voluntary, a drastic difference from the post-1949 period.[5] Moreover, regulators' early involvement in the operations of the Shanghai industry mainly engaged in content censorship and placed restrictions on the distribution and marketing of foreign films (mainly from Hollywood and Europe).[6]

Zhu continues to illuminate how political interventions directly and indirectly impacted the film industries and their personnel. Chapters 3 and 4 cover the formation of a newly centralized socialist film industry and the attendant film culture, which was guided by the leading party's ideology and the nation's shifting diplomatic relationships during the Cold War. Zhu shows how the Chinese government's admiration for the Soviet Union's approach to the nationalization of industry, including the film industry, resulted in tightened control of socialist political ideology over film productions. This sharp ideological turn not only "painted the city of Shanghai red," to use Yomi Braester's phrase,[7] but also discontinued distribution and public screenings of Hollywood films. Zhu singles out two key characteristics of the socialist-era film industry. First, during the socialist period, domestic films were evaluated and criticized based on political motifs and stances. Second, the distribution and screening of foreign films in the People's Republic of

---

4. Zhu, *Hollywood in China*, 38.
5. Zhu, 25.
6. Zhu, 34.
7. Yomi Braester, *Painting the City Red: Chinese Cinema and the Urban Contract* (Durham, NC: Duke University Press, 2010).

China (PRC) were restricted due to changing diplomatic relations. Although barred from public screenings, Hollywood films were screened in small, private settings among the top leadership and for selected filmmakers to aid in the production of model opera films. Zhu focuses on the invisible yet continuing influence of Hollywood in China through the so-called internal reference films. Behind the screening of many "poisonous films"[8] stands the work of Shanghai Film Dubbing Studio, which includes translation and dubbing of those internal reference films.

Restrictions on imports of foreign TV and film productions gradually relaxed in the Reform era, which is the focus of chapters 5 and 6. Hollywood popular media and culture reentered public life through TV screens in the 1980s and occupied a substantial portion of the foreign film quota since 1994. In these two chapters, Zhu depicts the post-socialist Chinese film industry's continuous marketization process and the competing political and cultural powers involved. Zhu argues that turning to Hollywood was an inevitable step for regulators in the mid-1990s to reinvigorate the declining domestic film market.[9] The method to reignite China's domestic market through the importation of Hollywood films worked well, and the audience in China embraced these films. However, as Zhu observes, the popularity of Hollywood films raised serious concerns about potentially shrinking the market share of domestic films. As regulators presented more protective measures, Chinese filmmakers were eager to learn from Hollywood how to produce diversified popular films for local audiences.

Many film scholars[10] have linked the emergence of Chinese high-concept blockbusters, *dapian*, to Zhang Yimou's *Hero* (2002). In contrast, Zhu argues the model extends back to the 1995 productions; for instance, *Red Cherry* 红樱桃 (dir. Ye Daying 叶大鹰, 1995). In addition to analyzing

---

8. Zhu, *Hollywood in China*, 99.

9. Zhu, 133.

10. Yiman Wang, "Remade in China: Cinema with 'Chinese Elements' in the Dapian Age," in *The Oxford Handbook of Chinese Cinemas*; Yeh Davis, "Re-Nationalizing China's Film Industry," 37–51.

the rising popularity of blockbusters, Zhu also examines the evolution of the government's regulation of the film industry and a cultural policy that prioritized "soft power" and a media "going-out" strategy. As China grew to become the second-largest economy in the world, it became important to build a more robust infrastructure to facilitate film-market expansion. During this time, China and Hollywood successfully explored coproduction and coinvestment models, allowing both to reach their individual goals of cultural expansion and financial returns.

In chapters 7 and 8, Zhu presents the recent examples of popular films in the PRC market. Through case studies from two prominent mainstream film directors' work in the past decade, Zhu examines the box office success of the New Year films directed by Feng Xiaogang and "Lost in" series of comedy road films by Xu Zheng. After answering the question of why some films perform phenomenally in China's domestic market, Zhu tackles the more difficult question of the industry's largely failed attempts at "going out." Specifically, Zhu explores why mainland films are not popular in the United States.[11] Eventually, Zhu interprets the Xi-era global vision and cultural expansion through the "Belt and Road" initiative and the expansion of the film market in new regions of the world, compared with the Shaw Brothers in the 1920s and 1930s[12] and the ambition of Shaw in the 1960s.[13] The last part of the book captures the current tensions between the two states and the ramifications on the two film industries with narrowed market access and significantly reduced coproductions and collaborations, which all point to a final post-pandemic uncertainty.[14] Zhu ultimately views the temporary decoupling of Sino-Hollywood positively, as it will "allow for films of diverse style, politics and cultural persuasions to flourish."[15]

11. Zhu, *Hollywood in China*, 226.
12. Zhu, 236.
13. Zhu, 238.
14. Zhu, 254.
15. Zhu, 282.

Overall, the strength of *Hollywood in China* is that the author juxtaposes the history of Chinese film and Hollywood, offering a consistently comparative perspective and capturing a complex relationship that alternates between competition and cooperation. This ambitious and comprehensive project ends with the most recent entanglements of the two states in the global-political and cultural arena. The author's reference materials are detailed and include a wide range of Chinese- and English-language historical archival material, interviews, scholarly work, trade journals, etc. These are used to support extensive case studies of prominent film theories, filmmakers, and film productions. Zhu's work is certainly an update on the current research and adds a new conversation with the others in the field.

While the book is informative and equipped with great historical details on the two big film industries in the past century, some readers might find it beneficial to have more detailed elaboration of certain key concepts—for instance, "soft power" and the "Hollywood Way." After analyzing China's soft power strategy and ambitions to grow its global cultural influence through Hollywood's China-friendly films, one challenging question left unanswered is why this strategy remains unsuccessful. Though Zhu depicts a complicated history of what forces might make a Chinese or Hollywood film "popular" in China, the lingering question remains unsolved: Why do popular Chinese domestic films with high production quality only attract domestic and limited diasporic audiences? By comparison, "Cool Japan" and "Korean Wave" media have gained global attention in the past two decades. Additionally, when discussing soft power and global cultural image, Zhu largely leaves out the role of international award-winning arthouse films mostly by fifth-generation and six-generation filmmakers. Finally, the book mainly focuses on the theatrical distribution of films. It gives little space to the digital distribution of Hollywood films in China as the streaming services took off after the 2000s.

Overall, the book will benefit a readership interested in how sociopolitical changes and negotiations between regulators, filmmakers, film companies, and audience play out in the dynamic Sino-Hollywood relationship. The implications of (de)coupling the current two largest film industries in the world and the impact of it on cultural globalization deserve further discussion and attention.

The Cautionary Tale of Painting War Remembrance in China as a New Nationalism

Review of *China's Good War: How World War II Is Shaping a New Nationalism* by Rana Mitter, Belknap Press, 2020

FUWEI ZUO

Keywords: Second Sino-Japanese War, Chinese nationalism, memory politics, war remembrance

For more than seventy years after World War II, the Second Sino-Japanese War remains a lively topic in China. Rana Mitter, a prominent historian in Britain who specializes in modern Chinese history, has put forth his newest work on China's rediscovery of its wartime memory. His previous publications, *Forgotten Ally: China's World War II, 1937–1945*, and *China's War with Japan, 1937–1945: The Struggle for Survival*, all contributed to the arguably less-discussed Chinese war of resistance in the English-language literature. Mitter's newest work, *China's Good War: How World War II Is Shaping a New Nationalism*, portrays the postwar historiographical shift in China's war narrative and how memories contribute to the longevity of war legacies in modern Chinese policymaking, popular culture, and diplomatic strategies.

Mitter aims to analyze the modern discourse surrounding World War II in China, specifically the manifestations of the war in historiographical arguments, diplomacy, online communities, movies, and museums.[1] The author crafts firstly a historiographical chapter on the evolution of the Chinese-state narrative regarding the war. Mitter is correct that before the Deng era, the Chinese official and social memory of the Second Sino-Japanese War largely revolved around Communist leadership in war efforts, as many Chinese citizens would recall. The post-1980 shift in Chinese historiography allowed discussions of nationalist war efforts to enter the Chinese academy and society.[2] The necessity of a narrative shift partly explains China's memory vacuum in the immediate postwar era, a period in which European nations actively pursued reconciliation leading to a study of collective European trauma.

Mitter identifies the Chinese discourse on the war as having both liberal and restricted elements.[3] As the Chinese state eases its claim on war leadership, diversification of narrative occurs in both the Chinese public sphere and Internet communities. Mitter draws on an intriguing phenomenon called *guofen* 國粉 on the Chinese Internet where pro-Kuomintang (KMT or Chinese Nationalist Party), people are free to voice their admiration of the Nationalist Chinese government and its war efforts while constantly exchanging heated debates with pro-Chinese Communist Party (CCP) voices.[4] Such a relaxation of narrative is almost unthinkable in the pre-1980 atmosphere, and it also appears in the positive image given to the Chinese Nationalist forces on screen, with recent blockbusters like *The Flowers of War* (2011) and *The Eight Hundred* (2020) achieving overwhelming popularity among the Chinese public. In contrast, their receptions overseas have been much more limited, as non-European war narratives struggle to find markets in the Western film industry. Or, as Mitter puts it, prejudices and

---

1. Rana Mitter, *China's Good War: How World War II Is Shaping a New Nationalism* (Cambridge, MA: Belknap Press, 2020), 7.
2. Mitter, *China's Good War*, 15.
3. Mitter, 17–18.
4. Mitter, 145, 159–60.

limitations from the West in fact have prompted China to promote its war narratives.[5]

Since "new nationalism" constitutes the book's title, readers would expect it to be one of the structural backbones of Mitter's thesis. Ideally, new nationalism would entail a theoretical framework that introduces the evolution of nationalism in modern Chinese history, specifically the underlying old and existing Chinese nationalism. Instead, Mitter positions the Chinese framework of war remembrance—namely, its state-led mnemonic policies—into a competition of ideology between China and other liberal countries. It would be constructive if Mitter included a historiographical chapter to contextualize what is new about the new nationalism. Examples of earlier Chinese nationalist thought such as Sun Yat-sen's *Three Principles of the People*, which entailed characteristics of nationalism based on culture, ethnic identification, and polity, would provide an excellent historiographical evolutionary model for Mitter's take on Chinese nationalism.

Mitter does not view Chinese politics through an ethnonationalistic lens (Han nationalism), nor does he paint Chinese diplomatic actions as superior culturalism. The new nationalism he presents to the readership is akin to one of Eric Hobsbawm's theories, where right-wing European political parties fanned nationalistic sentiments among the populace to expand their social base.[6] Mitter applies this European practice to China and calls it *circuits of memory*,[7] where the Chinese state brought social memory of the Second Sino-Japanese War into the modern construction of Chinese nationhood and international identity. *Circuits of memory* make sense when considering the state-led national museum projects and the 2017 history textbook revisions regarding the war dates. However, by limiting the scope of the new nationalism solely to China's rediscovery of its wartime memories, Mitter's usage of the term is similar to James Townsend's 1996

---

5. Mitter, 158.
6. Roger D. Markwick and Nicholas Doumanis, "The Nationalization of the Masses," *Oxford Handbooks Online*, 2016, 4–5, https://doi.org/10.1093/oxfordhb/9780199695669.013.21.
7. Mitter, *China's Good War*, 14, 48.

prediction that China will wield its "assertive" nationalism on unresolved territorial claims,[8] along with the worsening potential for Sino-Japanese relations moving forward.

Furthermore, without a precise definition and comprehensive historiography, Mitter regrettably steers his discussion of new nationalism along the lines of geopolitics and regional power play. He repeatedly mentions China's assertive diplomacy in the Asia-Pacific region, referring to China's status as *daguo* 大國, a big country, in the speeches by President Xi and senior stateman Yang Jiechi.[9] While China is undoubtedly a *daguo* in size, this characterization should be seen within its historical frame. One might recall Deng Xiaoping's 1974 UN speech where the Chinese leader remarked that national sovereignties should be counted as equal regardless of their physical size, and China would not partake in hegemonism in its future international relations. Xi Jinping, too, made prominent mention in 2009 of China's noninterventionist practices in its international relations. In this book, Mitter presents mainly the hawkish responses of Chinese diplomacy in recent years without mentioning China's diplomatic restraint in unison. Hopefully, his observations about the new nationalism would invite further studies into Chinese war narratives, not limited to diplomatic actions, but also considering the Chinese social and political atmosphere at the time.

In terms of sources in his chapter on the Chinese historiography of postwar memories, Mitter has extensively consulted senior mainland historians on their positions over the post-1980 transition of the Chinese war narrative. However, Mitter provides scant evidence to back up some of his arguments. One notable example of such inadequacy lies in his analysis of the debate over the war's dates, where the Chinese government revised the starting point of the Second Sino-Japanese War to an earlier

---

8. James Townsend, "Chinese Nationalism," in *Chinese Nationalism*, ed. Jonathan Unger (New York: Routledge, 2015), 18–20.
9. Mitter, *China's Good War*, 88.

point. Mitter assumes that political pressures from northeast China potentially prompted the date revision, since the 1931 Japanese invasion of Manchuria remained a relatively isolated incident outside the traditional Chinese narrative of an "eight-year war of resistance."[10] Mitter backs up his argument simply by referring to his 2018 interview with a senior Chinese historian, yet he does not disclose the interview's context and the interviewee.[11] It is not clear why Mitter omits the name of the Chinese historian, since his interview reveals ample "backstage" information behind the potential political overreach into Chinese academic circles, and such a mysterious interview would naturally arouse questions from an informed readership.

Mitter favors the traditional "eight-year war" narrative by calling 1937 a more plausible starting date. He backs up his position with an argument that people in China and Japan did not perceive themselves to be at war before 1937, using the 1933 Tanggu Truce as a case in which hostilities came to a halt between the two states.[12] However, civilian publications in Republican China, starting from 1931, widely expressed Chinese civil outrage toward Japanese invasions.[13] Additionally, Mitter draws on the Western liberal studies of memory to guide how the Chinese memory framework should be understood. Viewing China through a liberal lens is fair as long as Mitter acknowledges the fundamental differences between the Chinese postwar memory and Western ones. An example of the above manifests in Mitter's criticism that mainland China refrains from commemorating Nationalist

---

10. Mitter, 84, 92.

11. Mitter, 71, 79, 100.

12. Mitter, 92.

13. See publications between 1931–1936: Chen Binhe 陳彬龢, *Dong Bei Yi Yong Jun* 東北義勇軍 [The brave volunteers of the Northeast], 1st ed., ebook. Reprint (Ri Ben Yan Jiu She, 日本研究社 [Japan Research Group], 1932), https://taiwanebook.ncl.edu.tw/zh-tw/book/NCL-9910010553; Hua Zhenzhong 華振中 and Zhu Bokang朱伯康, *Shi Jiu Lu Jun Kang Ri Xue Zhan Shi* 十九路軍抗日血戰史 [The history of the 19th Route Army's bloody battles], 1st ed., ebook. Reprint (Shanghai: Shen Zhou Guo Guang She 神州國光社, 1947), https://taiwanebook.ncl.edu.tw/zh-tw/book/NCL-003150271.

casualties during the Chinese Civil War.[14] One can articulate a comparison between the Chinese Civil War with the American one, that under the liberal frameworks of memory, the American Confederacy on the losing side still enjoys a degree of commemoration in the southern United States. Such an argument is problematic since, firstly, the KMT authorities in Taiwan, too, made no commemoration of the Communist casualties. Secondly, the Chinese Civil War is technically still ongoing, with no formal end of hostilities ever signed by either belligerent. Therefore, by Mitter's reasoning, if a liberal commemoration of all Chinese Civil War casualties is ever to occur, it can only transpire with the war's final cessation.

If all the arguments mentioned earlier produce room for more profound scholarly debates on the Chinese memory of the war, Mitter stresses a few other ideas that would surely attract heated contestations. In the book's concluding chapter, the author firmly paints China's war narrative as a quest for international prestige, virtue by sacrifice, and leverage in the policymaking in the Asia-Pacific region.[15] Mitter points out the debatable nature of aligning the Nanking Massacre with the Holocaust in that such events are incomparable since the Japanese slaughter of Chinese civilians and soldiers did not equate to an "attempted genocide."[16] Mitter subjectively suggests that Chinese diplomacy is both "inept" and "often clumsy,"[17] that China's current quest to seek restorative justice as a victim during the war is unwelcome, since contemporary China has become immensely powerful.[18] Mitter's last argument is deeply problematic because he conflates two separate concepts into one—namely, the strength of a nation and the quest for justice. Should China, Korea, and other Asian countries ravaged by imperial Japan forgo undue justice simply due to the improvement in their conditions? Or, should nations give up seeking justice for war crimes because a number

---

14. Mitter, *China's Good War*, 211, 257.
15. Mitter, 218, 236.
16. Mitter, 238.
17. Mitter, 221, 239.
18. Mitter, 242.

of their people have "moved on" through the passage of time? Mitter's argument might be applicable if the International Military Tribunal for the Far East (IMTFE) had thoroughly carried out the sentences given to Japanese war criminals, yet this was not the case; by 1956, the majority of convicted Class A Japanese war criminals were released from Sugamo prison.[19] Although Mitter attempts to steer the revision of Chinese war narratives in the age-old rhetorical direction of the "China threat," his political spectatorship about unresolved justice will likely face a cold reception in China and among many of its neighbors.

Overall, Mitter's book serves as a fine general introduction to how the Chinese state attempts to direct war narratives and how war memories became integral in modern Chinese society for readers interested in the relatively new field of Chinese postwar memories. Mitter's work is valuable because he portrays aspects of Chinese social memory of the war similar to the liberal Western memory model. The divide between pro-KMT and pro-CCP voices in Chinese online communities, private and state-initiated commemorations, and diversified narratives in television and film regarding the war all separate China from stereotypical stigmas that perceive it as a state where only official narratives are allowed. However, the readership should exercise caution where Mitter's political commentaries are concerned. It is fair to compare the Chinese model of memory and its Western liberal counterparts, yet Mitter's overuse of "new nationalism" as his response to Chinese policymaking is damaging for readers who wish to observe the Chinese perspective on war memory with a degree of objective distance. While war memories in Europe may have seen amalgamation into a collective European trauma, for China and its neighbors, the trauma of Japanese militarism lingers on, unresolved.

19. David Cohen and Yuma Totani, *The Tokyo War Crimes Tribunal: Law, History, and Jurisprudence* (Cambridge: Cambridge University Press, 2018), 53.

Tracking American Political Currents

Review of *White Identity Politics* by Ashley Jardina, Cambridge University Press, 2019, and *Fox Populism: Branding Conservatism as Working Class* by Reece Peck, Cambridge University Press, 2019

DAVID GURNEY

It is both tired cliché and profound truth when we say that we are living through a tumultuous and complicated period of American political unrest. Antagonistic tribalism is at the heart of this, with the body politic seeming intent on splitting itself down the middle with a line of demarcation separating "us" from the evils of "them." The struggle to make sense of this is vital, and Ashley Jardina's *White Identity Politics* and Reece Peck's *Fox Populism: Branding Conservatism as Working Clas*s are very real attempts to help us better comprehend some of the forces that have propelled us here.

Jardina is a political scientist intent on understanding the role of race in American political life, and through her research, she finds a unique and necessary perspective on the topic. She contends that our dominant lens on racism—one that has largely presupposed that the bias of whites against racial and ethnic minority groups, especially blacks, stems from outwardly directed animus—is insufficient for viewing the full extent of the racial dynamics most profoundly impacting our political discourse. *White Identity Politics* argues that by focusing on the disparagement and hatred toward others, we miss the critical element of white identity as a

https://doi.org/10.3998/gs.3570

galvanizing mechanism for in-group solidarity. While we have more readily recognized solidarity among disadvantaged sociodemographic groups
in society, the role of in-group favoritism among whites has largely been
ignored.

The book contends that a major part of this blind spot has been in how
we discuss whiteness in general. Whiteness has been allowed to exist nearly
invisibly, as a racial group beyond race, consistently positioned at the lead of
the racial hierarchy while rarely acknowledging this in meaningful ways while
more overt expressions of racial bias have become seemingly less tolerable.
Jardina asserts that this situation has shifted in recent years due to two major
events. One is that immigration and birthrates have increased racial and ethnic
diversity in the United States, with demographers prognosticating that whites
will cease to make up a numerical majority in the near future. The other is the
election of Barack Obama in 2008, which signaled to many the unsure future
of white leadership. These events have made white cultural dominance feel less
certain, leaving many whites in a position to have their white identification
"activated" in the face of what they perceive to be their group's loss of status.

Jardina's research relies heavily on data obtained through the American
National Election Studies Time Series and a few other key initiatives meant
to gather information about the demographic makeup of the American elec-
torate, which she mines well to provide a unique and crucial perspective
on what constitutes white identity and consciousness and how these interconnected concepts operate. Her project feels both obvious and surprising,
starting with her overarching move to redirect the attention away from the
outward animosity and/or systemic biases that exist against minority groups
to the attitudes and dispositions that exist among whites that lead to the
favoring of their own identity group above all others. It seems likely that
many will bristle at this similarly to the way that some feminist critics and
scholars have taken issue with strains of gender studies that redirect attention from the injustices of patriarchal culture toward the masculine identities that perpetuate and benefit from that culture. These sorts of reactions
are necessary in that part of what gender studies and critical race studies

seek to repair is the deficit in serious attention given to subaltern groups and those with subaltern identities. However, the move toward focusing on the subaltern exclusively within scholarship looking to understand the workings of identity leaves us with another kind of knowledge gap. Although we may succeed in deterring the most obvious sorts of destructive behaviors when we call out overt bias, there are other, relatively more hidden (or ignorable/rationalizable) ways for bias to persist and have wide-ranging impacts. Jardina's project shines a light on why this situation will persist if we do not find ways to interrogate and better apprehend the role of white identification and solidarity within society.

Some of the most unexpected analysis comes in the demographic breakdown that she develops based upon her sifting of survey data. Jardina finds that white identity aligns less with party affiliation, economic status, or regionality than many would assume. The most significant correlations she finds based upon her evidence are that those who are most likely to have strong white identity and consciousness "are lower in education, higher in authoritarianism, and with greater levels of [social dominance orientation]," with those last two being "personality traits" that are in some sense harmonious.[1] The authoritarian leaning describes those who favor stricter, centralized leadership be maintained, which makes sense if one feels they are a member of a dominant group that benefits from the status quo. The social-dominance orientation is one in which those possessing it want their in-group to have privileges beyond those in any out-group. While other factors such as being older, living in rural areas, and having a traditionally blue-collar profession also correlate to some extent, these are much weaker than the relationships with education and the aforementioned personality traits or inclinations. This helps to adjust our view of how racial division operates and provides us a pathway to understanding how widespread white identification is.

---

1. Ashley Jardina, *White Identity Politics* (Cambridge: Cambridge University Press, 2019), 115.

The fifth chapter of *White Identity Politics*, "The Contents and Contours of Whiteness," is perhaps the most illuminating of the entire book, as it clearly articulates what constitutes whiteness as those most connected with the identity experience it. This is at the core of the intervention that this project makes. Jardina finds that white identification aligns strongly with a vision of Americanness that emphasizes being a natural-born citizen, a speaker of English, a believer in Christianity, and, of course, a person with white skin. More importantly, she finds that those with strong white racial identity can both recognize their own privileged status while they simultaneously report feeling aggrieved about what they perceive to be a narrowing of opportunities for themselves and fellow whites in addition to other disadvantages for their race, much of them centering around concern over "reverse racism." Even among whites who do not embrace white identity, such rejection is often couched in a rejection of race considerations altogether (aka "color blindness"), which also leads to a perpetuation of racial inequality through the maintenance of the status quo of white dominance.

Jardina goes on to trace how white racial identity aligns with political stances, including immigration policy, federal aid programs, and preferred presidential candidates. Her analysis reveals that while some of the positions of those with strong white identity—anti-immigration, anti-globalization, proisolationism—fall along lines that could readily be seen as having racial animus undergirding them, there are many positions regarding social welfare, affirmative action, and other policies benefiting racial minorities that do not align so well. Jardina explains this as indication of in-group favoritism being a much stronger force than out-group animosity, even if many of its practical effects have the same results. This is readily apparent in her breakdown of white identity's role in the 2012 and 2016 United States presidential elections.

While whiteness is not overtly the core concern of Reece Peck's *Fox Populism: Branding Conservatism as Working Class*, it is very much imbricated in what he ultimately uncovers and argues. As the title of the book suggests, its focus is the rise and impact of Fox News as a twenty-four-hour news channel, particularly in its stylistic approach to television journalism and the conservative political identity that the style manifests. Rather than looking directly at how white identification informs American politics of the early twenty-first century as Jardina does, Peck provides in-depth critical explication of how Fox News has played a major role in cementing a marriage of tabloid tactics and populist rhetoric and performance as key modes of address for contemporary American conservatism.

Fox Populism builds its argument on rich textual and contextual analysis of Fox News programming and marketing, as well as its foundations, especially in the pre–Fox News careers of Rupert Murdoch, Roger Ailes, and Bill O'Reilly. In mining that, the book establishes the Fox style as one rooted in a strong blend of tabloid sensationalism and populist politics that can be traced back to the earliest days of those key players. Peck also situates this within the broader post-network television moment in which cable channels were looking for ways to better court niche audiences. In this assessment, Fox News is not so much an unforeseeable game changer as it is a logical development of various forces that were already in motion within the changing US political landscape and media ecology.

Much of the analysis of actual Fox News programming content is anchored in their coverage of the financial crisis of 2007–2008 and its aftermath, which is a particularly opportune window onto its style. For one thing, Fox had, by then, been firmly established as the ratings leader among twenty-four-hour news competitors. For another, the crisis overlapped the campaign, election, and early presidency of Barack Obama. Perhaps most importantly, it was a moment in which socioeconomic status seemed poised momentarily to be a prime pivot point within political discourse. This was a crisis rooted in the actions of financial institutions and the various regulatory (and deregulatory) moves made by elected officials and governmental entities, and that is largely where the initial blame and concern was directed. However, as Peck's research bears out, Fox played a critical role in shifting this narrative from one of "Wall Street greed and corporate malfeasance to one centered around fiscal policy and the national debt, taxpayer victimization, and the 'sweetheart' benefits of public-sector workers."[2] The details of how that occurred reveal a lot about how Fox discursively constructs their visions of conservatism and conservatives themselves.

The third and fourth chapters provide the most thorough case for how the Fox News brand functions in terms of projecting a working-class identity for itself and its audience. One major component here is the movement from constructing social class as a primarily cultural phenomenon rather than an economic one. Peck challenges the work of others, such as Thomas Frank, who have advanced the notion that conservative populism sows a false consciousness that has the working class misrecognize the interests of the wealthy elite as their own by obscuring economic concerns with cultural ones. Wanting to undo that binary, Peck demonstrates how the cultural dimensions of working classness are just as salient as economic ones and how Fox has so opportunely tapped into this misreading of the impact of cultural class distinction on the part of the political Left. This shift of emphasis from wealth status to the taste culture of class allows for the displacement of populist anti-elitism against the rich onto an anti-elitism that rejects education, expertise, and nonworking-class taste as the real signifiers of the elite. It also laid the foundation for the rise of what Peck describes as the rhetoric of "entrepreneurial producerism" that is best encapsulated in the framing of the wealthy as "job creators." When tied together with cultural populism, this paints a picture of social hierarchy as a story of moral imperative, with those on top always deserving their position due to their superior work ethic while those on the bottom (at least as they are seen by others) suffering their fates as the result of their individual failings of ingenuity.

By providing shifts in critical perspective, both these books complicate the ways in which we view the contemporary political terrain of the United States. While the Trump presidency is not central to building the arguments in either of these projects, it looms large as the situation which

---

2. Reece Peck, *Fox Populism: Branding Conservatism as Working Class* (Cambridge: Cambridge University Press, 2019), 1.

these arguments attempt to illuminate and is what both these books have to inevitably address toward their closings. *Fox Populism* titles its conclusion as "Trump Populism," which makes great sense given how its preceding argument almost makes Trump's success with the Republican base seem inevitable, with his own style so readily fitting, and amplifying, the style of Bill O'Reilly, Sean Hannity, and Fox News more generally. Trump's performances at his rallies and events exhibit the same blend of tabloid hyperbole and conservative populist "common sense" tactics that long defined Fox News' style, and Trump's career in real estate, at least in how he presents it, positions him very well within a social hierarchy calibrated by the logic of entrepreneurial producerism.

Perversely enough, as Peck charts, Trump's rise and success came amid a moment in which Fox was incrementally pulling back from its tabloid populist brand and attempting to present a more respectable and less exclusively working-class image. This was also a time in which rival news outlets, on television, radio, and online, had risen to rival some of the stranglehold that Fox seemed to have had on conservative politics. A short postscript offers an attempt at mapping the resulting shift toward a different sort of conservatism, labeled as "alt-right," and championed by some of those rival outlets, particularly the website Breitbart News. This strain of conservatism has more "unapologetically built on white identity politics and anti-feminism and that emphasized economic nationalist positions over conservatism's traditional free-trade internationalism."[3] To anyone familiar with the events surrounding the 2016 US presidential election, the results of this swing are apparent in the tight connection forged between Trump and Breitbart, and especially its former executive chairman Steve Bannon.

While there are strains of the same tabloid populism at play, Peck questions whether the shift away from working-class alignment and the more transparent appeal to white identity evident in the alt-right style will resonate as strongly as the Fox style had. Even though Jardina's *White Identity*

---

3. Peck, *Fox Populism*, 239.

Politics offers the vantage of a different angle more founded in racial identity politics, it arrives at a similarly uncertain assessment of the longer-term viability of Trumpism and the alt-right. The more overt appeal to white identity and consciousness apparent in Trump and his relationship with alt-right voices and news outlets was couched enough in, or given cover by, his anti-immigration, anti-globalization, and pro-isolationism messaging that it worked effectively for the 2016 election, but, over time, this approach could ultimately become too blatant to maintain its dominance among those who are less comfortable with naked racial animus. This speaks to the general uncertainty surrounding the 2020 US presidential election and its aftermath. From what we can see now, there is still clearly some utility to the rhetorical appeals of working-class populism, tabloid sensationalism, and a preservation of the status quo that benefits those with strong white identity, but it may be getting stretched to a point where the group that it serves is no longer big enough to sustain its momentum. However these dynamics ultimately play out, it behooves us all as scholars and citizens to expand our understanding of these factors, which have brought us to this period of turmoil, and it seems essential that political strategists and activists include consideration of them as they look toward the future of democratic discourse in the United States.

CONTRIBUTORS

Kenneth Paul Tan is a tenured professor of politics, film, and cultural studies at Hong Kong Baptist University (HKBU), which hired him under its Talent100 initiative in February 2021. He teaches and conducts interdisciplinary research at the Academy of Film, the Department of Journalism, the Department of Government and International Studies, and the Smart Society Lab. His books include *Movies to Save Our World: Imagining Poverty, Inequality and Environmental Destruction in the 21st Century* (Penguin, 2022); *Singapore's First Year of COVID-19: Public Health, Immigration, the Neoliberal State, and Authoritarian Populism* (Palgrave Macmillan, 2022); *Singapore: Identity, Brand, Power* (Cambridge University Press, 2018); *Governing Global-City Singapore: Legacies and Futures after Lee Kuan Yew* (Routledge, 2017); *Cinema and Television in Singapore: Resistance in One Dimension* (Brill, 2008); and *Renaissance Singapore? Economy, Culture, and Politics* (NUS Press, 2007).

Louis Menand is the Lee Simpkins Family Professor of Arts and Sciences and the Anne T. and Robert M. Bass Professor of English at Harvard. His books include *The Metaphysical Club*, which won the Pulitzer Prize in History, the Francis Parkman Prize from the Society of American Historians, and the Heartland Prize from the *Chicago Tribune*. He has been associate editor of the *New Republic* (1986–1987), an editor at the *New Yorker* (1993–1994), and contributing editor of the *New York Review of Books* (1994–2001). Since 2001, he has been a staff writer at the *New Yorker*, which he began writing for in 1991. In 2016, he was awarded the National Humanities Medal by President Barack Obama.

Contributors

Martha Bayles has published hundreds of essays and reviews in literature, the arts, popular culture, and media. A fellow at the Institute for Advanced Studies in Culture at the University of Virginia, she is currently at work on a book about free speech in the digital age. She is a contributing editor to the *American Purpose*, film and TV critic for the *Claremont Review of Books*, and the author of two books: *Hole in Our Soul: The Loss of Beauty and Meaning in American Popular Music* (Free Press, 1994) and *Through a Screen Darkly: Popular Culture, Public Diplomacy, and America's Image Abroad* (Yale, 2014). Since 2003 she has taught interdisciplinary seminars in humanities and political science at Boston College.

Ilona Jurkonytė, PhD, is a Lithuania-born film and moving image researcher and curator. From 2015 to 2019, she was the Vanier Scholar at Concordia University, Canada. Her background is in philosophy, communication studies, and art theory. Ilona's academic research and curatorial practice inform each other. Her work critically examines tensions between notions of the national and transnational in moving-image production and circulation, as well as their geo- and hydropolitical implications. Ilona's research interests include the analysis of military imaging and mapping, film, and media participation in hydrospatial politics. Ilona was cofounder (2007), managing director (2007–2012), and artistic director (2007–2019) of Kaunas International Film Festival (Lithuania), which was instrumental in rescuing *Romuva*, the oldest cinema theater in Lithuania. Before starting her PhD studies, she worked as film program curator at the Contemporary Art Centre Cinema (Vilnius, Lithuania). Ilona also collaborates on creating single- and multiple-channel audiovisual works.

Tarik Cyril Amar is a historian of eastern Europe, especially Russia and Ukraine. His research interests include World War Two, the Cold War, authoritarianism, nationalism, memory, film, television, politics, and culture. He has published *The Paradox of Ukrainian Lviv: A Borderland City between Nazis, Stalinists, and Nationalists* (Cornell University Press, 2015).

His *Screening the Invisible Front: Spy Heroes and Popular Culture in the Soviet Union and Eastern Europe during the Cold War* is forthcoming with Oxford University Press. He also comments on current affairs in various outlets and @TarikCyrilAmar on Twitter. His website is tarikcyrilamar.com.

Noit Banai, PhD, Columbia University, is an art historian and critic who specializes in modern and contemporary art in a global context, with a particular focus on conditions of migration, exile, diaspora, border regimes, and statelessness. Before joining Hong Kong Baptist University, she served as professor of contemporary art in the Department of Art History at the University of Vienna and lecturer of modern and contemporary art at Tufts University/School of the Museum of Fine Arts, Boston. Banai is author of *Yves Klein* (Reaktion Books, 2014) and *Being a Border* (Paper Visual Arts, 2021). Her articles appear in journals such as *Third Text, Stedelijk Studies, Public Culture, Performing Arts Journal,* and *Texte zur Kunst.* Her essays have been published internationally by Sternberg Press, Centre Georges Pompidou, Museum Moderner Kunst Stiftung Ludwig Wien, Musée d'art moderne de la ville de Paris, documenta, Kontakt Collection, Americas Society, and the Bronx Museum of the Arts. Banai's current project, "Stateless: Jewish Artist Refugees in Shanghai, Hong Kong, and Singapore, 1933–1950," is connected to analyses on the function of borders as a constitutive aspect of the modernity/coloniality matrix, the plight of refugees, and the way in which "states of exception" have given rise to the multidirectional production of official and nonofficial memory cultures.

Kenny K. K. Ng is associate professor at the Academy of Film at Hong Kong Baptist University. His published books include *The Lost Geopoetic Horizon of Li Jieren: The Crisis of Writing Chengdu in Revolutionary China* (Brill, 2015); *Indiescape Hong Kong: Interviews and Essays,* coauthored (Typesetter Publishing, 2018) [Chinese]; and *Yesterday, Today, Tomorrow: Hong Kong Cinema with Sino-links in Politics, Art, and Tradition* (Chunghwa Book Co., 2021) [Chinese]. He has published widely in the fields of comparative literature, Chinese literary and cultural studies, cinema and visual culture in the United States, the United Kingdom, Europe, Hong Kong, Taiwan, and China. His ongoing book projects concern censorship and visual cultural politics in Cold War Hong Kong, China, and Asia, Cantophone cinema history, and left-wing cosmopolitanism.

Marina Kaneti is assistant professor in International Affairs at the Lee Kuan Yew School of Public Policy, National University of Singapore. She works at the intersections of visual politics and political theory and explores questions of geopolitics and global governance, including world order, migration, climate change, and heritage.

Yongli Li is an assistant professor of Chinese Studies at the College of the Holy Cross in Massachusetts. Dr. Li earned her PhD in the Department of East Asian Languages and Cultural Studies at the University of California, Santa Barbara. Her research focuses on the representation of Shanghai in contemporary film and media productions. She has coauthored chapters in edited volumes, including *Film Marketing into the Twenty-First Century* (2015) and *From Networks to Netflix: A Guide to Changing Channels* (2022). She is interested in Chinese film history, urban studies, and media policy.

David Gurney is an associate professor of media studies at Texas A&M University-Corpus Christi. His work on comedy and digital media forms, aesthetics, and politics has appeared in *Velvet Light Trap, Convergence: The International Journal of Research into New Media Technologies*, and *Flow*, as well as a number of anthologies. He is currently working on a project dealing with the circulation and popularization of extremist political ideologies and conspiracy theories via digital media.

www.ingramcontent.com/pod-product-compliance
Lightning Source LLC
Chambersburg PA
CBHW071712170526
45165CB00005B/1982